'Complete this programme as directed for 21 days.
The reading and actions will take less than an hour
a day – really! The results of your gene reprogramming
will benefit you for the rest of your life – as long
as you remain on the Primal path.'

MARK SISSON

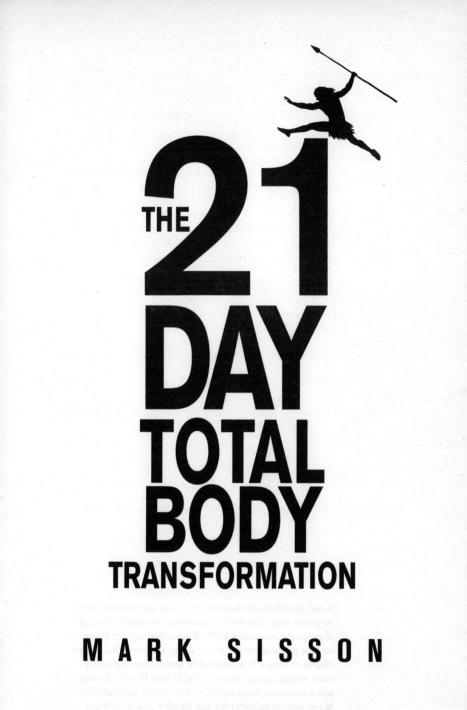

THE 21 DAY TOTAL BODY TRANSFORMATION

MARK SISSON

Vermilion
LONDON

1 3 5 7 9 10 8 6 4 2

This edition published in 2012 by Vermilion, an imprint of Ebury Publishing
A Random House Group company

First published in the USA by Primal Nutrition in 2009

The Random House Group Limited Reg. No. 954009

Addresses for companies within the Random House Group can be found at
www.randomhouse.co.uk

The Random House Group Limited supports The Forest Stewardship Council
(FSC®), the leading international forest certification organisation. Our books
carrying the FSC label are printed on FSC® certified paper. FSC is the only forest
certification scheme endorsed by the leading environmental organisations,
including Greenpeace. Our paper procurement policy can be found at
www.randomhouse.co.uk/environment

Printed and bound by CPI Group (UK) Ltd, Croydon, CR0 4YY

ISBN 9780091947842

To buy books by your favourite authors and register for offers visit
www.randomhouse.co.uk

The ideas, concept[...] [...]e used
for educational [...] [...] that
author and publi[...] [...] is this
book intended to [...] [...] eat any
disease, condition, [...] [...] any diet
or exercise progran[...] [...]ramme,
you receive full me[...] [...]r claim
no responsibility [...] [...]aused
or alleged [...] [...]
applic[...]

CONTENTS

INTRODUCTION

My first book, *The Primal Blueprint* (published in June 2012), took three years of research and writing and over 30 years of immersion into the world of diet, fitness, performance nutrition and elite athletics. I was honoured by the resulting critical acclaim and sales popularity, as it climbed to the number two best-seller overall on amazon.com in March 2010. Personal satisfaction aside, the popularity of *The Primal Blueprint*, my website, marksdailyapple.com, and the primal/paleo/hunter-gatherer movement in general sends the clear message that people are tired of flawed Conventional Wisdom and are ready to embrace the lessons of our past when striving to lead a healthy, happy, active life.

'Going Primal' offers an appealing alternative to the distorted messages conveyed by Conventional Wisdom – that we have little control over the ageing process, or our genetic predispositions to health problems and excess body fat. Instead, you can actually reprogramme some of your genes, press the 'reset' button on the runaway decline in human health in the modern world and reconnect with your own recipe for optimal health, ideal body composition and longevity that has been moulded by two million years of human evolution. In fact, as you will soon discover, it is your birthright to be lean, strong, fit, healthy and happy. I'm here to take you through the steps to regain that birthright. After *The Primal Blueprint* was published, I started receiving recurring back-handed compliments along the lines of, 'Wow – really comprehensive work Mark, great detail and excellent research. But … what exactly do I do now? What are the precise steps I can take to start living Primally today?'

The 21-day Total Body Transformation is the answer to that question. It's a practical, action-orientated guide on how to eat, exercise and live Primally – a 'cut to the chase' resource to make a smooth and quick transition into a Primal lifestyle. First, we will cover eight Key Concepts – 'things you need to know' to succeed. These Key Concepts represent the most important day-to-day elements of the Primal Blueprint and address some of the common questions posed by followers of the Primal Blueprint. Once you are armed with the knowledge and understanding of these Key Concepts, you can proceed with confidence and focus to tackle the five general Action Items – 'things you need to do'. Finally, you will jump into the 21-day Challenge – 'time to get to it!'. This is a step-by-step journey of daily challenges (categorised as Diet, Exercise or Lifestyle challenges) relating to the Action Items, with corresponding journal exercises.

> The Primal Blueprint *is about getting the greatest health and fitness benefits you can with the least amount of pain, suffering and sacrifice.*

So, when I ask you to throw away a large percentage of the food that currently occupies your fridge and kitchen – staples that probably have sustained you for your entire life – you will be buoyed by a clear understanding about how these dietary shifts will profoundly improve your metabolism and health. Or when I explain why you really don't need to spend that much time exercising, you won't be inclined to doubt me (or try to sneak in some 'extra credit' workouts!)

Unlike many other diet and exercise 'programmes', the Primal Blueprint philosophy offers tremendous flexibility for personal preference, and even the hedonistic enjoyment of comfortable modern life. You can be sure that I walk my talk, but I'm no ascetic or tightly wound fitness freak. As I described in detail in *The Primal Blueprint*, I've been there and done that during my career as an elite marathon runner and iron man triathlete. Today, my goal is to look super fit without having to follow an exhaustive, time-consuming exercise regimen, enjoy my meals without the slightest hint of deprivation or restriction and essentially neutralise the ageing process by engaging in lifestyle behaviours that promote optimal gene expression. I want you to experience that kind of freedom and empowerment as well.

A 21-DAY TRANSFORMATION THAT WILL LAST THE REST OF YOUR LIFE

Your genes are extremely sensitive to environmental influences, many of which you control directly, so you can make great progress towards reversing years of adverse lifestyle habits in just 21 days. By eating and exercising Primally, you will change from a sugar-dependent, fat-storing organism that constantly battles hunger, illness, depression and weight gain, into what I call a 'fat-burning beast' who burns stored body fat day and night, at exercise and at rest, as your primary energy source. By recalibrating delicate hormonal processes that have been thrown out of balance by a hectic modern life, you will be able to manage stress smoothly, avoid burnout, enjoy elevated immune function and high energy levels all day long, fall asleep easily each evening and awaken refreshed each morning.

Twenty-one days represents a benchmark in the gene repro-gramming process, and, not coincidentally, is also believed by many experts to be the length of time required to eliminate old habits and replace them with new ones. Make a sincere effort to follow this programme for 21 days and you will be transformed for life – as long as you remain on the Primal path.

If you visit marksdailyapple.com and check out the Success Stories link, you will see that profound changes in body composition happen quickly when you go Primal. A few of my favourite Success Stories are also presented here. It's reasonable to expect a reduction of 1.4 to 3 kg (3 to 7 lb) of excess body fat in your first 21 days, and to continue at that rate safely until you reach your ideal body composition. I'm confident that the benefits you will experience will promote a transformation in not only how you eat and exercise for the rest of your life, but also in your beliefs about diet, exercise, ageing, immune function, prescription medication and numerous other elements of Conventional Wisdom that have been surreptitiously compromising your health for decades. One of the most common testimonials on marksdailyapple is 'I've lost 50 pounds effortlessly, but it feels like I've also lost 1,000 pounds off my shoulders – because I know I can live like this for the rest of my life!'

While you'll hopefully experience an exceptional first 21 days, it's possible that you'll have some difficulty to start with, depending on your particular history and lifestyle circumstances. If you've spent years eating the typical Western diet marked by heavy consumption of processed carbohydrates, sugars and certain unhealthy fats, and if you have been immersed in what I call a 'Chronic Cardio' exercise pattern, you may struggle with occasional swings of hunger and energy for the first 7 to 21 days of eating Primally. Have no fear, these symptoms will dissipate with each passing day as you gradually repair some of the metabolic damage from the typical Western diet habits, normalise your insulin levels and reprogramme your genes to burn stored body fat for energy.

SHOOT – JUST FOR KICKS!

Consider taking a 'Before' photograph at the outset of your 21-day Transformation. This is not about winning a trip to Hawaii by starving yourself, exercising to exhaustion for three weeks, and snapping an amazing 'After' photo. Rather, it's recording for posterity the starting point of a long term lifestyle transformation.

This exercise should be fun. Please don't feel any frustration or negativity about your starting point. Simply grab your mobile phone camera, pose in front of a mirror with minimal clothing, and snap a full-body shot. That way, you can take and store the photo in complete privacy if you wish!

In 21 days, you will likely be down several pounds of excess body fat (if you indeed wish to lose body fat), a jeans size or two, and you may have toned or added muscle depending on your commitment to exercise. Your physique changes will certainly be reflected in the mirror, but that should not be your central focus. Instead, emphasise the enjoyment of the process, and allow results to happen at a natural and comfortable pace. In three months' time, or a year's time, your body composition improvement may be dramatic enough that you feel compelled to snap an 'After' photograph and submit it for publication at marksdailyapple.com!

TAKING RESPONSIBILITY

The 21-day Total Body Transformation is characterised by flexibility and personal empowerment. Hence, I hesitate to use words like 'programme', 'regimen' or 'diet' when talking about the 21-day Total Body Transformation. Really, this journey is about understanding the behaviours that promote optimal gene expression, and taking responsibility for the consequences of your day to day lifestyle choices. When you embrace this mentality, you will be able to skirt the all too common phenomenon of diving enthusiastically into a programme, regimen or diet, becoming fixated on a specific result, and then losing momentum for various reasons: the 'programmes' are too difficult, impractical or physically and mentally stressful to sustain long term, you don't achieve the result you expected and you don't have any fun in the process. Worst of all, you are likely to experience a decline in health despite devoted efforts to do the right thing by Conventional Wisdom.

Your 21-Day Total Body Transformation will be characterised by flexibility and personal empowerment. Your Primal efforts must be fun, energising, and easy to maintain at all times, otherwise you are destined to fail. What are the best workouts to do, or foods to eat? Whatever you enjoy the most!

> *Your Primal efforts must be fun, energising and easy to maintain at all times, otherwise you are destined to fail.*

There is no greater feeling of empowerment than when you begin to comprehend how much influence you have over your health, fitness and well-being. Once you realise that your genes respond to environmental signals that you largely create, you are no longer at the mercy of your parents' legacy, your doctor's nebulous warnings or the tremendous momentum against health and balance in a hectic modern life. Everything changes from the time you first 'own' the Key Concept that you can influence gene expression on a day-to-day basis. In many cases, you can choose which genes to flip on and which to flip off through your food and activity choices!

This is a profound responsibility to reflect upon. We live in a world of such abundant choice and freedom that we can direct gene expression away from health and still not suffer any penalty in the traditional 'survival of the fittest' evolutionary sense (today, unfit humans are not eaten by predators and, furthermore, are able to reproduce freely!) For two million years, humans were subjected to unimaginably severe selection pressures and so adapted and thrived to enjoy a position at the top of the food chain. Since the advent of civilisation 10,000 years ago, the typical evolutionary selection pressure ceased to the extent that we are literally 'devolving' from the pinnacle of human health represented by our hunter-gatherer ancestors.

I strongly support taking advantage of free choice, but I'd like you to reflect for a moment on the obligation you have to yourself, your loved ones and the planet to take good care of yourself. Today, cloaked in the veneer of affluence and rampant consumerism, there is far too much unnecessary, expensive and totally preventable pain and suffering caused by poor health practices and unconscious lifestyle choices. This programme is about you making informed choices with personal responsibility and empowerment!

WHOM TO TRUST? HOW ABOUT YOU!

I must admit that some of the Primal Blueprint Key Concepts are hotly contested, with respected scientists and health professionals passionately defending their life's work on both sides of the debate. It can be quite disconcerting to absorb conflicting advice when determining the best course of action for your health. I don't want to force dogma down your throat and urge that you simply take my word for it with so much on the line – particularly since I'm not an accredited scientist or doctor. Instead, I will present you with my interpretation of a variety of scientific, medical and anecdotal data relating to the Key Concepts, and let you decide for yourself. But the anecdotal evidence of the success of my 21-day Transformation approach is there: I have had the pleasure of seeing thousands of people lose fat, increase energy and regain excellent health by following the Primal Blueprint.

> *Regardless of your starting point, past failures or bad luck with familial genes, you can turn things around quickly – starting with your next meal and next workout. Your genes expect you to be lean, strong, energetic and healthy.*

First, you might want to take stock of how things are working for you right now. For example, are you someone who enjoys engaging in chronic exercise – always balancing on the edge of burnout, illness and injury? Do you enjoy strictly controlling fat intake in the name of weight control so avoiding some of the most delicious foods on the planet (steak, eggs, bacon, butter, macadamia nuts, avocados, etc.)? Does portion control and the hassle and regimentation of trying to balance calories each day work effectively for you? Do you already enjoy optimal health, fitness and body composition, and have you managed to avoid the overstress/burnout syndrome that is epidemic in our hectic modern world, or could things be a little – or a whole lot – better?

It's clear that even today's most devoted health and fitness enthusiasts struggle with weight control, recurring fatigue and minor illness, chronic pain in joints and muscles, increased risk factors for lifestyle-related diseases and an ageing process that is vastly more accelerated and debilitating than it has to be. What's also apparent from recent scientific advances is how readily and efficiently our bodies embrace behaviours that promote optimal gene expression. Regardless of your starting point, past failures or bad luck with familial genes predisposing you to excess body fat and other health problems, you can turn things around quickly and build momentum for lifestyle transformation – starting with your next meal and next workout. You can literally recreate, rebuild and renew your body using the Primal Blueprint principles. Based on two million years of human evolution, your genes want – and expect – you to be lean, strong, energetic and healthy.

> *Regardless of your starting point, past failures or bad luck with familial genes, you can turn things around quickly – starting with your next meal and next workout. Your genes expect you to be lean, strong, energetic and healthy.*

I have great respect and interest in science, medicine, epigenetics, evolutionary biology and exercise physiology, but I default to personal experience whenever I'm conflicted about what is the best course of action to promote my health. I strongly encourage you to do the same. If you aren't having fun at workouts, then whatever you're doing is wrong for you. If you're not enjoying meals, then your diet is literally unhealthy. One caveat here: I aim to be healthy, happy and energetic and I want to achieve that with the least amount of pain, suffering, sacrifice, discipline, calorie counting and portion control possible. I will make the assumption that you do too.

THE 21-DAY TOTAL BODY TRANSFORMATION KEY CONCEPTS AND ACTION ITEMS

The eight Key Concepts create the framework around which you will smoothly move into a custom-designed Primal lifestyle. Once you understand these concepts, you can implement the five Action Items with confidence – dialling in your eating, exercise, sleep and play for the rest of your life.

KEY CONCEPTS

1. **Yes, You Really Can Reprogramme Your Genes:** More than just determining your fixed heritable traits, genes are responsible for continually directing the production of the proteins that control how your body functions every second. Genes turn on or off only in response to signals they receive from the environment surrounding them – signals that you provide based on the foods you eat, the types of exercise you do (or don't do), sun exposure, and even the air you breathe.

2. **The Clues to Optimal Gene Expression are Found in Evolution:** Two million years of selection pressure and harsh environmental circumstances created the perfect genetic recipe for human health and longevity. Our genes expect us to be lean, fit and healthy by modelling the lifestyle behaviours and diets of our hunter-gatherer ancestors – even in the context of a hectic modern life. Plants and animals (meat/fish/poultry/eggs, vegetables, fruits, nuts and seeds)

should comprise the entirety of the human diet, with allowances for the moderate intake of certain modern foods. As for exercise forms and frequency, less is often more.

3. **There is No Requirement for Dietary Carbohydrates in Human Nutrition:** Conventional Wisdom's grain-based, low-fat diet has artificially created a sugar and carbohydrate-based metabolism that you've been stuck in, and suffering from, for your entire life. Going Primal shifts you into the fat-based, all-day energy metabolism that has supported human survival for two million years. This is the most liberating aspect of Primal living.

4. **80 Per Cent of Your Body Composition Success is Determined by How You Eat:** Many modern foods (even ones you thought were healthy) are causing you to gain weight and get sick. Moderating insulin production by giving up grains, sugars and pulses and lowering inflammation by eliminating harmful man-made fats, will promote efficient reduction of excess body fat, effortless maintenance of ideal body composition, increased daily energy levels, decreased risk for illness and optimal function of various other hormone systems (stress, appetite, immune, metabolic, sleep, thyroid, etc.).

5. **Grains are Totally Unnecessary (And So are Pulses, For That Matter):** The centrepiece of the typical Western diet offers minimal nutritional value, promotes fat storage by raising insulin and contains anti-nutrients that promote inflammation, compromise digestion and often interfere with immune function. There is no good reason to make grains (or pulses, for that matter) any part of your diet unless you want a cheap source of calories that easily converts to sugar.

6. **Saturated Fat and Cholesterol are Not Your Enemy:** The Conventional Wisdom story about heart disease is only validated when you eat lots of sugar and refined carbohydrates. Cholesterol is one of the body's most vital molecules. Saturated fat is our preferred fuel. The true heart disease risk factors – oxidation and inflammation – are driven strongly by polyunsaturated fats, simple sugars, excess insulin production and stress. Limiting processed carbohydrates and eating more high-quality fats and whole foods (including saturated animal fat) can promote health, weight management and reduced risk of heart disease.

7. **Exercise is Ineffective for Weight Management:** Burning calories through exercise has little influence on your ability to achieve and maintain ideal body composition. When you depend on carbohydrate (sugar) as your primary fuel, exercise simply stimulates increased appetite and calorie intake. Chronic exercise patterns inhibit fat metabolism, break down lean muscle tissue, and lead to fatigue, injury and burnout.

8. **Maximum Fitness Gains Can Be Achieved in Minimal Time with High-Intensity Workouts:** Regular brief, intense strength training sessions and occasional all out sprints promote optimal gene expression and broad athletic competency. You enjoy more benefits in a fraction of the time spent doing the chronic exercise advocated by Conventional Wisdom.

ACTION ITEMS

1. **Eliminate the Typical Western Diet Foods:** Out with the undesirable foods that promote weight gain and chronic health problems.

2. **Shop, Cook and Dine Primally:** Restock your kitchen with Primal foods, and implement winning strategies for shopping, meal preparation, dining out and snacking.

3. **Make The Healthiest Choices Across the Spectrum:** Understand how to make the best choices in the categories of meat, fish, poultry, eggs, vegetables, fruits, nuts and seeds, fats and oils, foods allowed in moderation such as dairy, and occasional sensible indulgences.

4. **Exercise Primally – Move, Lift and Sprint!:** Pursue broad athletic competency with an intuitive blend of workouts honouring the three Primal Blueprint Fitness laws (Move Frequently at a Slow Pace, Lift Heavy Things, and Sprint Once in a While).

5. **Slow Life Down:** Take the time to enjoy simple pleasures such as 'slow food' over industrialised food, balanced instead of chronic exercise, focused work habits instead of multitasking, interpersonal relationships over social media, calm, relaxing evenings instead of excessive artificial light and digital stimulation, and plenty of time for play, sun exposure, rest and relaxation.

THE 21-DAY TOTAL BODY TRANSFORMATION – SNEAK PREVIEW

Here's a quick overview of the lifestyle changes you will make during your 21-day Transformation, and beyond. The way you proceed with your 21-day Transformation is up to you. If you are the deliberate, analytical type, feel free to read the entire book before you embark on your first kitchen clear out or Primal workout. If you are the enthusiastic, fast-action type, you can implement lifestyle changes as you make your way through the book, building some momentum immediately with the following 'Out with the Old' and 'In With the New' sneak previews. We will describe everything over the course of the book, but the essence of Primal living is not complicated. In fact, it's about reducing the complexity of modern life and adapting the simple lifestyle practices of your hunter-gatherer ancestors into today's world as best you can. It's all about doing what your genes expect you to do to be lean and healthy.

OUT WITH THE OLD

1. **Grains, sugars, sweetened beverages:** Processed carbo-hydrates drive excess insulin production, which can lead to lifelong insidious weight gain. Even if you don't have excess body fat concerns, a high insulin-producing diet promotes systemic inflammation, fatigue and burnout. Eliminating grains from your diet may be beneficial as they also contain 'anti-nutrients' which may cause health problems beyond just gaining weight.

2. **Industrial 'PUFA' oils:** Trans and partially hydrogenated fats (from heavily processed snack or frozen foods), deep-fried menu items (from fast-food restaurants), assorted packaged snacks and baked goods (crisps, crackers, biscuits, etc.), margarine-type spreads and bottled vegeta-ble oils (rapeseed, corn, safflower, etc.) promote oxidation and inflammation, setting the stage for cancer and heart disease.

3. **Beans and other pulses:** Beans, lentils, peas and soya products contain anti-nutrients that compromise diges-tion, immune function and general health. The fibre so highly touted in beans is mostly indigestible, and the carbodydrate content in all pulses is high enough to warrant cutting or eliminating them in the interest of moderating insulin production.

4. **Dairy:** Most dairy that you find in typical stores is unhealthy for everyone (whether or not you are lactose intolerant). The good news is that butter and double cream are the preferable forms of dairy, provided you don't have lactose intolerance issues.

5. **Chronic exercise:** Workouts that are too hard, too long and done too frequently with insufficient recovery lead to burnout and failed weight-loss efforts. Reject the notion that reaching a consistent level of mileage, hours or workout frequency is the key to fitness.

6. **Sedentary patterns:** Prolonged sedentary periods (commuting, desk jobs and digital entertainment) promote fat storage, elevated cardiovascular disease risk, joint pain, muscle weakness and diminished energy and focus on peak performance tasks.

7. **Poor sleep habits:** Excess artificial light and digital stimulation in the evening disturb the optimal flow of sleep and stress hormones, compromising health, fitness, weight management and longevity.

8. **Sombre, spartan approach to lifestyle transformation:** No more calorie counting, portion control, rigid meal timing and menu choices, guilt or bingeing cycles. No regimented workout schedule or predetermined mileage, time or rep standards to attain without regard to daily fluctuations in energy, motivation and performance levels. No struggling or suffering in the name of health and fitness! Going Primal can truly feel easy, effortless and natural – once you break free from diehard old habits.

IN WITH THE NEW

1. **Primal foods:** Meat, fish, poultry, eggs, vegetables, fruit, nuts and seeds, high-quality fats, a moderate intake of high-fat dairy products and supplemental carbs (for heavy exercisers and growing youth), and occasional sensible indulgences such as red wine and dark chocolate.

2. **Primal eating philosophy:** Enjoy tremendous freedom and flexibility to choose your favourite foods and recipes within the incredibly broad Primal Blueprint guidelines. Eat to your heart's content with full awareness and appreciation of natural hunger and satiety cycles. Indulge sensibly with a clear conscience that it's okay to enjoy life!

3. **Increase daily movement:** Make a concerted effort to engage in more general daily movement (neighbourhood strolls, using stairs instead of lifts, spontaneous play sessions, walk breaks at work, etc.). Conduct regular low-level (easy) aerobic workouts at appropriate heart rates, and take frequent movement breaks when engaged in prolonged sedentary tasks.

4. **Brief, intense workouts:** This is a centrepiece for optimal gene expression in muscles, heart and lungs, and is essential to maintain high energy, anti-ageing, and broad athletic competency. Go harder but less frequently and for less duration. Thirty-minute strength workouts or 15-minute sprint workouts are plenty – any longer is probably too much for most people.

5. **Calming evening rituals:** After dark, minimise exposure to artificial light and digital stimulation, and wind things down with calming endeavours (e.g. strolling, reading, socialising).

6. **Fun approach to lifestyle transformation:** Explore exciting new foods, recipes and spontaneous, intuitive Primal eating practices. Exercise for energy and fun, and avoid overtraining. Tap into your youthful spirit by taking short breaks and grand outings. Power down your hyperconnectivity inclinations and appreciate the simple pleasures of family, friends and personal reflective time. Realise that being healthy and fit – even super-fit – does not have to involve suffering or deprivation, and can actually be fun!

WHAT TO EXPECT WHEN YOU'RE EXPECTING ... TO GO PRIMAL!

- **Anti-Ageing:** A 2011 study published by the Cooper Institute in Dallas suggested that one's fitness level – represented by how fast you can complete a 1.6 km (1 mile) run – is an excellent predictor of longevity. Primal Blueprint Fitness workouts will build strength, speed and endurance safely and quickly. Over time, fitness improvements will facilitate improved psychological health, further counteracting the ageing process as it is perceived today.

- **Appetite:** As your genes redirect your cells to derive more energy from fat and depend less on glucose, you will be freed from the need to eat frequent high-carb meals and snacks to continually prop up sagging blood glucose levels. Your appetite will 'self-regulate' to the point that you will tend not to overeat any more. Eating Primally also enables you to easily engage in Intermittent Fasting (IF) – both spontaneous or structured – to boost immune function, cellular repair and fat metabolism.

- **Blood Markers:** You can expect significantly lower triglyceride and 'bad' (LDL) cholesterol levels, higher 'good' (HDL) cholesterol levels, normalised blood sugar, healthier blood pressure and improvements in other critical blood test markers, in as little as 21 days.

- **Body Fat:** Expect to lose excess body fat at a rate of 1.8 to 3.6 kg (4 to 8 lb) per month until you achieve your ideal body composition. This happens incidentally when you switch to Primal eating, even if you've had extreme difficulty losing weight by the conventional calories in/calories out methodology. If, in rare circumstances, you struggle to attain this rate of fat loss, the Primal Leap programme (primalblueprint.com) will enable you to pinpoint fat reduction with greater accuracy.

- **Cravings:** As you train yourself to easily burn fat more and depend less on glucose (blood sugar) for energy, your cravings for sweets will lessen. You just won't 'have' to have them so often. Similarly, as you remove processed foods and grain-based products from your diet you will tend not to seek out commonly craved salty foods.

- **Digestion**: Eliminating processed foods foreign to your genetic make up will quickly help alleviate symptoms of digestive dysfunction, even lifelong 'issues' that may seem like normal by-products of stressful daily life, genetic frailties or ageing. You may even repair damaged digestive tissue. Allergies, asthma, inflammation – anything with '-itis' is fair game – will minimise or disappear in a matter of days or weeks.

- **Drug-Free:** Experiencing rapid improvements in blood markers and disease symptoms can enable you to progress towards an important goal of eliminating your reliance on prescription medications – something your doctor will support when your blood markers return to normal range.

- **Energy**: When your body changes to deriving most of your energy from stored body fat, you won't be subject to the blood glucose swings and burnout cycles that happen with the typical Western diet. Instead, you will notice – in 21 days or less – increased and more stabilised daily energy levels, even if you find yourself inadvertently skipping meals or 'forgetting' to eat.

- **Immunity**: Moderating production of cortisol (a primary catabolic stress hormone) and reducing levels of glucose and insulin in your blood will help your immune system function optimally. You will probably get sick less often and recover more quickly if you happen to get run down.

- **Measurements**: Your clothes will fit looser as you decrease body fat, systemic inflammation and the water retention and bloating that accompanies it. You will see the most results in your behind, hips, thighs and waist – typically the primary storage areas for fat.

- **Muscle Mass:** You can increase or sculpt muscle while dropping fat – or maintain muscle mass if you already have as much as you need. In contrast, the Conventional Wisdom approach of chronic exercise and high-carbohydrate eating commonly results in muscle catabolism (breakdown) in order to deal with wildly fluctuating blood glucose levels.
- **Sleep:** Aligning your lifestyle with your circadian rhythm will enable you to go to sleep easily, sleep soundly and awaken naturally (no alarm) each morning, refreshed and energised.
- **Stress:** Living Primally and rejecting the typical Western diet/chronic exercise pattern will regulate your body's stress response system, helping you to avoid the fatigue, burnout, disease and dysfunction that are driven by a hectic modern life.
- **Total Fitness:** Primal Blueprint Fitness involves full-body, functional exercises that develop broad athletic competency and a balanced physique. This allows you to pursue a variety of fitness and athletic goals without the risk of overtraining and injury common with narrowly focused programmes.
- **Various Other Health Markers:** Directing optimal gene expression improves bone density, glucose tolerance, insulin sensitivity, blood pressure, hormone balance and many other benefits ... including LGN.*

* LGN = Looking Good Naked!

KEY CONCEPTS
THINGS YOU NEED TO KNOW

KEY CONCEPT 1:
YES, YOU REALLY CAN REPROGRAMME YOUR GENES

Inside each of your cells is a DNA 'recipe' – a set of general instructions for how to build a lean, fit, happy productive human being. I say 'general', because how things actually play out in your life is a function of activating or deactivating thousands of very specific genes (which are subsets of the DNA) on a day to day basis.

Genes are commonly viewed as fixed traits that you inherit from your parents such as hair and eye colour, height, body type and predispositions to alcoholism, flat feet, rheumatoid arthritis, shyness, wide hips and so forth. You have minimal influence over heritable traits such as these, but you can directly influence genes involved in muscle development, body fat storage, inflammation and many other aspects of general health and longevity. Your ability to influence this gene expression is the very foundation of the Primal Blueprint.

Many thousands of genes are constantly at work throughout your life directing cellular functions: messaging protein molecules involved in building and repairing muscle and organ tissue; and repairing, regenerating and sometimes even destroying your cells based on the signals they get from your lifestyle behaviours. You can control these environmental signals through the foods you eat, the workouts you conduct, the sleeping and lifestyle habits you engage in and even the medication you take.

Genes can be viewed as an assortment of 'on/off' switches for building the protein molecules that influence every element of body function and structure. Turn on 'good' gene switches and you build muscle, increase fat burning or knock out an invading virus. Turn on 'bad' gene switches and you might experience inflammation and indigestion, or, over the long term, you may develop obesity, heart disease and cancer.

> *The genetic recipe for a strong, fit, healthy human exists in almost all of us. It's our 'factory setting' at birth.*

GENE EXPRESSION IN PRACTICE: OTTO AND EWALD

Otto and Ewald are German twins who possess identical copies of the genes their parents gave them. They achieved notoriety in scientific circles as an ideal case study for how environment influences gene expression. Otto was in training for long-distance running, while Ewald was a competitor in field events (discus, shot put, hammer throw) which call for brief, explosive bursts of power. Otto's low-intensity endurance workouts trained his muscles to process oxygen more efficiently, but partly deactivated the genes that trigger protein synthesis and increased muscle size. Ewald's high-intensity training increased gene activity involved in protein synthesis, so his muscles grew larger and more capable of brief, explosive efforts.

The moral of the Otto and Ewald story is that your final body shape and health status is not predetermined at birth. Your parents gave you a range of possible outcomes based on their own genes and the genes of their immediate forebears. Even if you inherited a tendency to store excess body fat around your hips like mum and Aunt Lucy, or possess one of the BRCA gene variants that increase your risk of breast cancer, good news is around the corner. A great deal of your so-called predispositions for chronic health problems and lifestyle-related disease can be eliminated or largely neutralised by sending the right signals to your basic *Homo sapiens* genetic profile. It just might take some people a little more effort or diligence than others, but knowledge is truly power when it comes to the control you have over your health destiny.

To really grasp this Key Concept, it's essential to recognise the difference between the particulars of your unique familial genes, and the basic human genes that we all share. We all build muscle and bone, burn and store fat or combat germs the same way, using the same biochemical pathways in our human genes. It's just the degree to which we do these things that varies among individuals. Some of us – thanks to mum and dad – build muscle a little faster and better than others. Some of us seem to burn off fat with far greater ease. Some of us have genes that make us more prone to getting cancer or heart disease. But in most, if not all, of these cases, you have significant power to affect whether genes get switched on or off. You just have to know which behaviours or which foods affect which genes. Luckily, discoveries in human evolution coupled with recent advances in mapping the human genome provide many of the clues.

Here's a big problem though: your genes have strong expectations to receive a specific and narrow range of signals from you, but they don't really know or care if you make 'good' or 'bad' lifestyle choices. These myriad on/off switches are hard wired from millions of years of evolutionary selection pressure to first and foremost keep you alive until you are old enough to reproduce. Hate to break it to you, but your genes will pursue this mission without taking into account the effect upon your long term-health. It's fair to assume that your genes prefer the path of least resistance, and are always ready and waiting to build a strong lean, fit, healthy, happy human. However, they will easily respond to signals that promote systemic inflammation or the excessive storage of fat if you choose to send those signals instead.

When you conduct an all-out sprint workout, your genes stimulate a pulse of anabolic hormones, enabling your body to adapt and grow stronger for your next sprint session. That's a good thing. On the other hand, when you mismanage your genes with poor dietary habits or chronic exercise patterns, you will likely suffer from obesity (through the chronic overproduction of insulin), fatigue (poor sleep habits disturbing optimal hormone balance) and systemic inflammation and burnout (chronic production of 'fight or flight' hormones in the face of unrelenting environmental stressors).

When you get type 2 diabetes, it's not necessarily a sign of defective genes: in fact, it's an example of your human genes (with some familial influence) doing what they think is appropriate to protect you from you having too much sugar in your bloodstream. However, abusing this lifesaving mechanism is definitely a bad thing for your long-term health. With this in mind, ironically, everyone is naturally 'predisposed' to developing type 2 diabetes if they send their genes the wrong signals enough times. The good news is that no one has to develop type 2 diabetes. It is the most preventable of all diseases – *if you send your genes the right signals.*

Your 'factory setting' at birth is to be the fittest, healthiest person you can possibly be – to realise your own genetic potential. It's important to recognise that 'results may vary'. Olympic athletes and magazine cover models represent optimal expression of their specific physical genetic attributes, but the hard wired limitations of your familial genes might be enough to preclude you from ever winning a gold medal. Accepting this reality, it's critical to settle for nothing less than superior health, maximum longevity, and avoid the devastating pattern of decline and disease that is endemic to modern life. Realising your genetic potential is as simple as knowing which switches to flip.

CHASE-ING PAUL ... AND MATT DAMON!

Paul and Chase Sheaffer of Pensacola, FL are Primal enthusiasts and identical twins. Their respective journeys to regaining their health are particularly compelling examples of the power that we all have to reprogramme our genes. Paul and Chase both became clinically obese at 5'9" (175cm) tall and around 300 pounds (136 kg), and suffered from assorted health problems as a consequence of sedentary, Standard American Diet (SAD) lifestyles. Would it be fair to say that the twins had bad luck with their familial genes?

In going from over 300 pounds to 150 (68 kg) in a matter of 15 months of Primal living (see Paul's success story on page 64), Paul demonstrated that while he may be predisposed to obesity from a high insulin-generating diet (as most of us are), he also has profound genetic gifts that enabled him to progress extremely quickly from a couch potato to a competitive martial artist! Inspired by Paul's success, Chase starting transitioning to a Primal lifestyle in 2011, and has also enjoyed spectacular results, dropping 70 pounds (31 kg) in the first five months.

Digging deeper into such a remarkable success story, it turns out that Paul and Chase's parents are both extraordinary athletes – dad was a professional gymnast, and mum a professional ballerina! Clearly, Paul and Chase possess exceptional familial genes – either exceptionally good, or exceptionally bad, depending on the signals each twin sends his genes in daily life. When you make the choice to promote optimal gene expression, you can actualise your gifts and nullify, or at least reduce, any genetic frailties and predispositions that will manifest if – and only if – you mismanage your genes.

YES, YOU REALLY CAN REPROGRAMME YOUR GENES: SUMMARY

- Genes are like 'on/off' switches for building protein molecules that influence every element of body function and structure.

- Genes are more than fixed heritable traits – they constantly direct the repair, rebuilding and regeneration of your cells.

- While some genetic functions are beyond our control (eye colour, skin pigment, etc.) we have tremendous influence over day to day operations through the environmental signals we send to our genes (diet, exercise and lifestyle choices).

- Genes strive to promote short-term health in response to all environmental signals, whether health-promoting long term or not (e.g. type 2 diabetes).

- Accept the limitations and predispositions of your familial genes, then focus on optimal expression of your human genes.

KEY CONCEPT 2:
THE CLUES TO OPTIMAL GENE EXPRESSION ARE FOUND IN EVOLUTION
(WITH VALIDATION FROM MODERN SCIENCE)

Noted evolutionary biologist Theodosius Dobzhansky once famously said, 'Nothing in biology makes sense except in the light of evolution.' Modern research in epigenetics and evolutionary biology confirms that we are genetically identical to our hunter-gatherer ancestors, a premise that frames today's popular Primal/

paleo/evolutionary-based eating, exercise and lifestyle movement. In order to discover which foods or behaviours can have the greatest impact on your genes, and thus advance you towards better health today, it's useful to look closely at human evolution for the past two million years.

The food, exercise habits and lifestyle behaviours that sustained evolution have shaped, moulded and supported the modern human genome (our complete collection of genetic material). Modelling the lifestyle behaviours of our hunter-gatherer ancestors (even in the realities of high-tech modern life) provides the ingredients to complete your own personal recipe for a lean, strong, fit, healthy, happy human. To help you visualise this, I've created a primal human role model affectionately known as Grok, and memorialised him in the Primal Blueprint logo.

Consider the familiar 'survival of the fittest' concept as it applies to your genes. Our ancestors, who were able to survive and repro-duce under unimaginably harsh environmental circumstances, refined and perfected the human genetic recipe. Those who were unable to adapt died out, and those lazy, slow, stupid, weak genetic attributes were lost forever. Today, your genes expect you to eat a certain way, engage in both extensive low-level movement and brief, high-intensity workouts, sleep in alignment with the rising and setting of the sun and so forth.

We may all rejoice in winning the survival of the fittest battle to enjoy our lofty perch at the top of the food chain, but we must not take the hard work and good fortune of our ancestors for granted. The recipe and the ingredients we need to maximise our health and well-being are right in front of us, but modern humans seem to disrespect and disregard the profound legacy of our ancestors.

Our ancestors, who were able to survive and reproduce under unimaginably harsh environmental circumstances, refined and perfected the human genetic recipe.

It might be hard to believe that we are exactly the same inside as the loin-clothed Grok from 10,000 years ago. With a few minor exceptions, we are identical to our ancestors in how we metabolise food, respond to exercise, cycle through sleep phases each night, absorb sunlight and deal with various other environmental influences. Oh, with one critical distinction: our ancestors from 10,000 years ago were stronger and healthier than most of us are today! Anthropological records from Grok's time show robust specimens with little or no heart disease, cancer, obesity, diabetes or autoimmune diseases. These hunter-gatherers lived on meat, fish, poultry, insects, eggs, nuts, plants and fruits and engaged in lots of varied physical movement in the thousands of years before agriculture.

Conversely, the initial generations of predominantly grain-based eaters – such as the Egyptians around 7,000 years ago – were significantly shorter, less muscular, had lower bone density and shorter life spans than their predecessors, and even exhibited dental decay. The decline in human health prompted by civilisation occurred because we departed from what our genes were accustomed to for the previous two million years: meat, fish, eggs, plants and constant activity. Instead, humans switched to eating brand-new agricultural foods (wheat, barley, peas and lentils were among the first cultivated crops) and adopted a less physically strenuous lifestyle. The abrupt lifestyle change of civilisation – what University of California, Los Angeles evolutionary biologist and Pulitzer Prize winner Dr Jared Diamond refers to as 'the worst mistake in the history of the human race' – dramatically altered which genes were switched on and which were turned off.

The genes of early civilised humans weren't adapted to these new foods, many of which contained toxins in the form of plant chemical defences. Members of these early agricultural societies often survived just long enough to have several children and to pass their genes on to the next generation, but you really couldn't say that these people thrived from a health standpoint. Even thousands of years later, with amazing technological, scientific and medical progress, we pretty much have the same situation today: people around the world – whether impoverished or wealthy – are simply surviving (from a health standpoint), not thriving.

Our genes still expect us to eat a higher fat diet. Our genes see many of these agricultural foods we take for granted as 'poisonous', because our guts haven't adapted. Our genes see an overabundance of sugar as toxic, and take dramatic steps to save us. Our genes see our lack of exercise, sleep and sunlight as problematic because we haven't adapted to being indoors, largely sedentary and blasted with excess artifical light.

> *Our genes still expect us to eat a higher fat diet; they still see agricultural foods (and modern foods such as sugar), as poisonous; they still see lack of sunlight and exercise as problematic. We haven't genetically adapted to modern life because there is no selection pressure in the civilised world.*

We simply don't give our genes what they evolved to expect from us, which begs the question: Why haven't we evolved to the point that we can thrive on these newer agricultural foods? Why are these foods still toxic thousands of years later? Couldn't we have evolved to eat these newer foods without penalty by now? The answer is all about selection pressure. Once civilisation took hold with permanent housing structures and relatively plentiful supplies of calories from agricultural efforts, the primary environmental selection pressures that had driven human evolution for the previous two million years by rewarding the genes of the best adapted were effectively eliminated. The constant threat of death by starvation or predator danger basically ended. Survival of the fittest has changed into survival of the weak, sickly, diabetic and arthritic. Consequently, the human race has been able to carry on, overpopulate the earth and see today's inhabitants of the most developed, affluent countries become the fattest, least fit population in the history of mankind.

Meet my pal Grok: he's genetically identical to you and me
– except stronger and fitter!

grok: Verb/gräk – to understand intuitively or by empathy.

LIVE LONG, DROP DEAD

You may have heard, accurately, that life expectancy was only around 33 in Grok's time, but this longevity statistic is much more impressive when you factor out predator danger, high infant mortality rates and routine infections and trauma that were commonly fatal in prehistoric times. Ten thousand years ago, anthropologists assert that it was not uncommon for humans who avoided rudimentary fatalities (eaten by lions, fatal infections from a scraped knee, etc.) to live six or seven decades in robust health – with no medical care or modern comforts of any kind.

Even with their lifelong struggle for food, shelter and safety, the 'maximum observed life span' in Grok's time was a mind-boggling 94! Impressive longevity is also enjoyed today among the last remaining pockets of primitive hunter-gatherer cultures on the globe, such as the Aché, Hadza, Hiwi and iKung. More than a quarter of today's Aché (pronounced Ah-chay) people in Paraguay make it to 70. Moreover, 73 per cent of Aché adults eventually die from accidents, and only 17 per cent from illness.

Meanwhile, the nearly octogenarian life expectancy in Western nations is tainted by a deplorable decline in health, vitality, fitness and productivity known affectionately as the ageing process. In contrast, Grok enjoyed truly exceptional health and fitness for his entire life, whether it ended early by misfortune, or whether he was able to go six or seven decades at full throttle. Grok actualises the motto I've recently come to embrace, 'Live long, drop dead!' There was simply no such thing as today's steady decline into old age, in some cases 20 years of slipping down the slope of declining health.

It may seem implausible to consider a bunch of freakishly fit, healthy and energetic primal humans roaming the earth, but it's the truth. Pre-civilised humans with genetic, physical or even psychological frailties were naturally and ruthlessly weeded out of the clan in a manner as routine as the rising and setting of the sun. Survival of the fittest ain't pretty, and thanks to technological progress, life today is easier, safer and more humane. Nevertheless, while no one wishes to return to the brutal realities of selection pressure, we must never take the legacy of human evolution for granted.

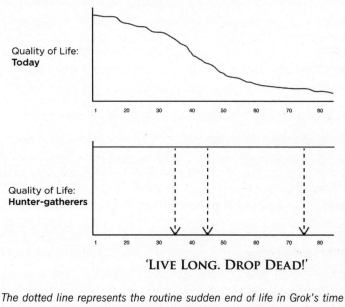

'LIVE LONG. DROP DEAD!'

The dotted line represents the routine sudden end of life in Grok's time – typically by accident or brief (and often minor by comparison today) illnesses – no steady decline into feeble old age.

The 21-Day Total Body Transformation is about making life-transforming *choices* based on solid scientific information. Knowing this, we can reject the complexity and gimmicks common to the latest diet or exercise craze, and look to our ancestors and evolutionary biology to discover optimal foods and movement patterns. This is not to suggest you must hunt down wild animals or forage for berries all day to be Primal aligned; it's about finding ways to use 21st century convenience, abundance and technology in ways that are consistent with our inherited biology. As we begin to drill down into what we should or shouldn't eat, our main criterion must be: will eating this food be consistent with my biology, or will it potentially cause problems? As we investigate the best forms of working out, we must ask ourselves: how does this exercise best take advantage of my genetic recipe for strength, power and endurance?

Due to the elimination of selection pressure and the insufficient time frame for evolutionary adaptation, modern research in epigenetics and evolutionary biology confirm that we are genetically identical to our hunter-gatherer ancestors from 10,000 years ago.

Today's exploding global population generates tremendous genetic diversity, but each of us still responds in a similar manner to environmental signals – food, workouts, sleep, sun exposure – that direct our own unique ideal genetic expression. The great news is that being healthy does not require extreme training, restrictive/obsessive dietary habits or a joyless, Spartan daily regimen. Thanks to the tribulations and triumphs of our ancestors, humans are hard wired to pursue a life of ease, contentment and happiness. We've adapted to a tremendous variety of foods and environmental circumstances in order to populate and thrive in all corners of the globe. Personal preference is going to have the strongest influence on lifestyle behaviours that promote optimal gene expression, as long as you operate in the broad framework that we will continue to outline in this section.

GROK'S LIFE LESSONS – THE ORIGINAL AFFLUENT SOCIETY

Grok's hunter-gatherer lifestyle reveals simple insights into how to be healthy. Regarding diet, the bulk of our ancestors' calories came from eating a variety of animal life (estimates range from 45–85 per cent, depending on geography), including insects, grubs, amphibians, birds, eggs, fish and shellfish, small mammals and some larger mammals. Berries and other fruit, leafy greens, primitive roots, shoots and other vegetation, nuts and seeds rounded out the hunter-gatherer calorific intake.

Absent from Grok's diet were the agricultural foods that appeared and became predominant over the past 7,000 years, as well as the modern foods that we have introduced over the past century. Loren Cordain, PhD, author of the *Paleo Diet*, claims that 71 per cent of daily calories in the SAD come from 'modern' foods that were entirely absent from the diet of our ancestors, and thus foreign to our hunter-gatherer genes: refined sugar products (to the tune of 70 kg/154 lb per person annually), grain foods (bread, corn, pasta, rice, etc.), pulses (beans, lentils, peanuts, peas, soya products), chemically altered trans and partially hydrogenated fats (deep-fried, frozen, packaged 'junk' foods), polyunsaturated fatty acids (vegetable and seed oils, packaged, baked and frozen goods – aka 'PUFA'), processed dairy products, meat from intensive factory farming laden with hormones, pesticides and antibiotics, and other packaged, refined, frozen, heavily processed fare that has been disastrous to human health.

71 per cent of daily calories in the typical Western diet come from 'modern' foods that were entirely absent from our ancestors' diets.

Another notable element of Grok's diet contrasts sharply with modern life – his calorific intake was wildly inconsistent. There was never any guarantee of where the next meal would come from, or whether Grok might become some other creature's next meal! Consequently, we adapted to the constant threat of starvation by becoming adept at storing energy for later use, and particularly adept at manufacturing glucose internally to fuel essential biological processes when dietary carbohydrates were scarce – which was often.

We store ingested calories in various ways: fat is stored as triglyleride in fat cells, carbohydrate is stored as glycogen in muscle and liver cells, and protein is stored as amino acids in muscle tissue. As we will discuss in detail on page 59, insulin is the key hormone that facilitates this storage. Efforts to consume regularly scheduled carbohydrate-based meals – three squares, six smalls, or other such modern gimmicks – actually do more harm than good by stimulating excessive insulin production and deactivating the genes that promote efficient fat burning. Our genes expect us to be healthy by often *not* eating regular meals!

Regarding lifestyle elements that promote optimal gene expression, the legacy of our ancestors has been seriously mischaracterised and distorted by our civilised values that lack appreciation for the redeeming qualities of primitive times. In the 17th century, English philosopher Thomas Hobbes popularised the notion that the lives of our pre-civilised ancestors were 'solitary, poor, nasty, brutish and short'. These beliefs continue to shape Conventional Wisdom even today, despite extensive research suggesting otherwise.

Anthropologist Marshall Sahlins has advanced a more dignified theory of hunter-gatherers as the 'original affluent society'. They wanted for little, met those desires on a daily basis and thereby enjoyed a level of affluence by the literal definition that was superior to the distorted values of today's consumerism culture. There was no need or desire to accumulate material wealth, vie for the prestige of having the biggest hut, hoard more food than their neighbour or engage in other stressful and unnecessary rat race battles.

> *Hunter-gatherers were the 'original affluent society'.*
> *They wanted for little, met those desires on a daily basis*
> *and enjoyed a life of extensive leisure time and rich*
> *social interaction.*

Our hunter-gatherer ancestors did just enough work to get by and enjoyed a life of extensive leisure time and rich social interaction. Estimates derived from modern-day hunter-gatherers suggest that Grok's routine probably consisted of 3 to 5 hours of procuring food, another few hours of chores relating to habitat, shelter and basic human needs, 10 hours of sleep and rest (probably bi-phasic – featuring a good afternoon siesta) and *six hours* of leisure time each day, consisting of play and family or group socialising. Rather than accumulating material goods, it would appear that ample leisure time (art, dance, music, play, sports and storytelling) is the true currency of the hunter-gatherer. Maybe that should be how we measure success today as well.

Hunter-gatherers' active lifestyle made them exceptionally fit, but they approached fitness from a 'necessity' perspective, with exercise flowing naturally from their basic needs: pick berries, seek out water, build shelters, or sprint from a predator. These exercise patterns contrast sharply with the chronic, narrowly focused exercise regimens common today. For example, humans aren't literally 'born to run', at least not daily. Our genes are wired for frequent walking, hiking, jogging (if one is quite fit) and general low to moderate intensity movement.

To be sure, we are adapted to perform occasional grand endurance feats, such as a persistence hunt. Check out YouTube for the documentary, *The Great Dance*, where Xo San Bushmen literally run an antelope to death over some four hours in the 49°C (120°F) heat of the Kalahari Desert in Botswana. Unlike today's mileage-obsessed marathon runners, however, our ancestors (and modern hunter-gatherers) balanced these awesome occasional efforts with extensive recovery periods.

Other simplicities of Grok's daily routine must be considered in light of the disastrous state of modern health. Aligning your sleep habits more closely with the rising and setting of the sun may not sound as exciting as consuming a variety of digital entertainment available 24/7, but it can enhance your health immeasurably. Carving time out of daily life to play, hang out and talk with friends, relax in the sunshine or take a relaxing evening stroll with your partner are activities that we commonly neglect in the name of productivity. However, embracing these seemingly trivial little diversions from your busy appointment calendar may help you to manage stress, depression, fatigue and high blood pressure better than a concoction of prescription medications. And your genes *expect* you to do all this to be healthy and fit.

THE CLUES TO OPTIMAL GENE EXPRESSION ARE FOUND IN EVOLUTION: SUMMARY

- Because civilisation ended selection pressure/evolution, we are genetically identical to Grok, a premise that frames the Primal Blueprint.

- The pinnacle of human physical evolution was reached 10,000 years ago, prior to civilisation. Grok was healthier, fitter and stronger than civilised humans today.

- Civilisation ushered in a decline in human health, due to eating, exercise and lifestyle patterns that compromised optimal gene expression for the first time. We've continued to mismanage our genes to become the fattest, sickest, least fit population in the history of mankind.

- Grok enjoyed excellent longevity (factoring out primitive risks/misfortunes) and quality of life. 'Live long, drop dead!' contrasts with the steady decline into old age today.

- Hunter-gatherer diet: plants and animals, wildly fluctuating calorific intake. In contrast, we eat 71 per cent of calories from genetically unfamiliar and offensive 'modern' foods.

- Hunter-gatherer exercise: fitness by necessity, which blends frequent low-intensity movement with occasional high-intensity strength and sprint efforts, and extensive recovery.

- Hunter-gatherer lifestyle: the 'original affluent society' which wanted for little, and fulfilled all needs daily. Sleep habits aligned with the sun. Lifestyle of slower pace, relaxation, play and socialising.

KEY CONCEPT 3:
THERE IS NO REQUIREMENT FOR DIETARY CARBOHYDRATES IN HUMAN NUTRITION
(YOUR BODY ACTUALLY PREFERS TO BURN STORED ENERGY!)

This might be the most important Key Concept in the book, so I really want you to get this: *your original 'factory setting' is to be an efficient fat-burning beast!* You are designed to derive most of your energy – minute by minute, day by day – from either the healthy fats in your meals, or from the body fat stored on your rear end, hips, thighs and waist. Your genes *expect* you to eat fats and to access stored fat for energy needs – both at rest and during low to medium intensity exercise.

Unfortunately, due to the misinterpretation of health and dietary science (and perhaps a little corporate profit incentive propaganda thrown in for good measure) that began a few decades ago, you were likely to be socialised into a typical Western diet at an early age. As 20th century lifestyle 'progress' took hold (eating food from factories instead of farms, driving instead of walking, etc.), and waistlines began to expand, Conventional Wisdom dispensed the flawed observation that eating fat makes you fat. We were convinced, erroneously, that carbohydrates (and the glucose and stored glycogen that they generate) are what our bodies prefer to burn, because they burn quickly and easily, and because certain vital organs like the brain cannot live without them.

Consequently, industrious food manufacturers revved up their machinery to pump out heavily processed carbohydrate foods and beverages. We cultivated and indulged our collective sweet tooth, and assuaged our health ambitions by eating low-fat, grain-based meals that we truly believed were nutritious and responsible for weight control. While the nuances of this issue might be debated by those in the wholegrain, low-fat camp today, one unassailable truth is that the typical Western diet of the past few generations has been an unmitigated disaster.

As we learnt in the examination of evolution and our ancestors' low-carb lifestyle, see page 27, a steady supply of dietary carbohydrates is entirely unnecessary, and counterproductive to our health. Yes, the brain needs a small amount of glucose to keep it running, but this and other essential metabolic functions involving glucose are easily handled by internal glucose manufacturing mechanisms that are hard wired into our genes. Your liver, if healthy, can make up to 130 grams of glucose – which it can also store as glycogen – on its own every day. This is more than enough to supply the brain and other organs, even if you never ate another carbohydrate. We only 'prefer' to burn glucose (from dietary carbohydrates) when it's present in large quantities, since excess glucose in the bloodstream is toxic.

In an effort to dispense with excess glucose as quickly as possible, your body burns it for immediate energy, stores it as glycogen in the muscles and liver or stores it as fat in your fat cells. Glucose is indeed a cheap and easy-burning source of fuel, but that doesn't mean that you should depend on dietary carbohydrates as a primary source of fuel. For one thing, operating a mostly sugar-burning engine (one that depends on fresh glucose all the time) will most likely lead to lifelong insidious weight gain unless you become a calorie counter and exercise junkie. Even big-time exercisers with exceptional genes who maintain lean physiques can suffer serious repercussions from being sugar burners, something we will discuss in the Exhaustion Epidemic section of Key Concept 4 on page 63. In addition, pumping too much glucose and insulin through your bloodstream over a lifetime promotes systemic inflammation, the catalyst for all manner of health conditions and serious disease.

As we will discuss in the next Key Concept, when you overstress your insulin response system and become 'insulin resistant', you are in big trouble. If glucose is not efficiently processed (either burnt, stored as glycogen, or stored as fat), it will damage protein molecules through a process known as glycation. In fact, the various health issues experienced by those with obesity and type 2 diabetes all generally relate to this glycation effect.

CHANGING FROM SUGAR BURNER TO FAT-BURNING BEAST

Like most people around the world, you probably grew up eating your fair share of high carbohydrate staples like whole grain and refined cereals, breads, pasta, fruit juices, sweetened drinks, potatoes, rice, rolls, biscuits, pastries, pies and all other manner of high carbohydrate foods. Even if some – or most – of your diet has been in the form of 'complex' carbs, this excessive and unnecessary intake of carbs has reprogrammed your genes from an early age to make you more dependent on a regular supply of dietary carbohydrate.

Operating a mostly sugar-burning engine will likely lead to lifelong insidious weight gain and systemic inflammation.

You are a 'sugar burner' if you are trapped in this carbohydrate dependency. If you are like most people, this high intake of carbs over time has elevated your insulin levels (see Key Concept 4) throughout the day, and resulted in excess calories (from all foods, not just carbs) to be stored as body fat. Over the years, chronically high levels of glucose and insulin in your blood have altered your fat cells so that you can't easily burn this plentiful source of stored energy. If your body is accustomed to burning carbs as fuel instead of stored fat, your brain will crave your usual fuel source, i.e. carbs, more often. This leads to a vicious circle of insidious weight gain over a lifetime, even if you exercise frequently.

We use the terms 'up-regulate' and 'down-regulate' to describe how gene pathways are activated or deactivated by the environmental signals they receive. With a diet very high in carbohydrates, and/or an exercise regimen lacking high-intensity workouts, genes down-regulate fat-burning processes, and up-regulate most of the enzyme systems and pathways that are involved in sugar burning and fat storage. The good news is that you can return your genes to their factory 'fat-burning beast' setting in a short time by eating, exercising and living Primally.

> *There is no requirement for dietary carbohydrate in human nitrition. Your body can manufacture glucose from protein and fats on demand to keep you focused and fuelled.*

Think about this: *there is no requirement for dietary carbohydrates in human nutrition.* Your body can manufacture glucose from proteins and fats on demand, and in amounts needed to keep your brain humming and energy levels stable, through a process known as *gluconeogenesis*. This elegant function happens in the liver where fats and proteins (either ingested or stored) are converted into glucose and pumped into the bloodstream to keep you focused and energised. Many experts believe that gluconeogenesis can supply you with up to 150 grams (600 calories' worth) of glucose per day if necessary. This is a significant amount, which might even cause you to second-guess the old wives' tale sugar-burning edict to consume regular meals in order to stabilise blood sugar.

You may have heard that gluconeogenesis is sometimes associated with the stress response, where hard-earned lean muscle tissue is stripped down when you exercise chronically, or endure a crisis in your personal life and run on empty for days on end. Gluconeogenesis can certainly be a less than optimal survival-based process in those types of situations (particularly before you become fat adapted and are more vulnerable to blood sugar dives), but it can also become a regular, daily useful tool in providing a steady supply of glucose (in the absence of dietary carbs) from ingested fats and proteins with no adverse effects. This is what happens when you choose to reprogramme your genes and cells to selectively derive most of your energy from stored body fat. This 'choice' is, of course, dependent upon the signals you send to your genes. To optimise your internal glucose manufacturing abilities, it's best to be fat adapted (eat Primally, moderate carb intake/insulin production), consume sufficient protein to preserve muscle tissue and avoid chronic exercise.

Entire civilisations have existed for ages on practically zero-carb diets. Mind you, these resilient people did not exist this way by choice, and I'm not suggesting that you should never eat carbs or that they are somehow inherently dangerous. I prefer to view carbohydrates as the 'elective' macronutrient, usually only needed in any appreciable amounts after heavy exercise when you want to replenish muscle glycogen (or perhaps also increase body fat, if you're a sumo wrestler or school football player). By simply cutting out the excess, undesirable carbs – sugary foods and beverages, refined and whole grains and pulses – and leaving in healthy carbs (e.g. those in vegetables, fruits and roots), you will arrive naturally at a biologically appropriate carbohydrate intake level that minimises your fat storage and sets you up to be an efficient fat burner for the rest of your life.

Before you breeze by this section and go on your merry way, I must assert that excess carbohydrates have a distinct, unique ability to ravage and metabolically disturb a person's body. If you're overweight, it's very likely that your carbohydrate metabolism is dysfunctional. You are probably insulin resistant to the extent that even a moderate amounts of carbs will inhibit healthy fat metabolism, promote systemic inflammation and suppress immune function. We will discuss these topics further in Key Concept 4 (see page 59) about insulin, and Key Concept 5 (see page 73) about the drawbacks of consuming anti-nutrients in grains and pulses. Of course, all or most of that dysfunction can be fixed over time by adhering to a Primal Blueprint lifestyle.

The truth is, fat and protein were the dominant macronutrients for two million years of human evolution, with carbohydrates only ascending to centre stage with the advent of agriculture around 7,000 years ago. Our ancestors often went days without anything to eat at all, yet they had to maintain their strength and mental focus until they could find more food. The lack of regular access to food and the scarcity of carbohydrates resulted in selection pressure to develop efficient pathways to access body fat for energy. This was the only way humans could survive day to day and generation to generation. Our genes are still programmed to rely on effective fat metabolism as our primary energy source – if we send them the right signals.

> *Our ancestors went for days without anything to eat, and carbohydrates were extremely scarce for two million years. The truth is, fat is the preferred fuel for human metabolism.*

Our movement patterns over millions of years (lots of low level fat-burning activity punctuated by brief bouts of intense sprinting or lifting) were such that we never needed to consume large amounts of glucose, or store large amounts of glycogen in our muscles and liver. It was predominantly stored fats and ketones (an energy source made in the liver as a by-product of fat metabolism – see page 48) that helped us to survive, evolve and thrive. Hey, there's another T-shirt slogan to go with 'Live Long, Drop Dead!': 'Survive, Evolve and Thrive!'

When you consider how ridiculously small the body's glycogen storage reservoirs are, you understand that it would have been impossible for us to survive as a species if glucose were truly the 'preferred' fuel. The liver, which is the main back-up glucose storage facility for the brain and other glucose-burning organs, can only store about 100 grams of glycogen, which is less than a day's worth. Your muscles can only hold another 350–500 grams. All told, that's barely enough to run for 90 minutes at a reasonable speed, as any marathoner who has 'hit the wall' and run out of glucose can attest.

Meanwhile, we have a virtually unlimited storage capacity for fat (close to a *million calories'* worth on some *Biggest Loser* contestants). Excessive glycogen storage has been unnecessary because we have tremendous capacity to store fat, to burn fat and to produce glucose and ketones through internal mechanisms we will discuss shortly. Besides, there just weren't that many carbs to eat for most of our two million years evolving.

THE INNER WORKINGS OF GLUCOSE

At any one time, the total amount of glucose dissolved in the bloodstream of a healthy non-diabetic adult is equivalent to only a teaspoonful (maybe five grams). Much more than that is toxic; much less than that and you pass out. That's not a big range for a so-called 'preferred' fuel, is it? Several studies have shown that under normal low activity conditions (i.e. at rest or walking) your body only burns about five grams of glucose an hour. And that's for people who aren't yet adapted to burn fat or ketones.

The brain is the major consumer of glucose, needing maybe 120 grams a day. Low-carb eating reduces the brain's glucose requirements considerably, and those who are eating a very low amount of carbs and are keto-adapted may only require about 30–50 grams of glucose per day to fuel the brain. Similarly, little to no glucose is required to fuel muscles when you exercise at 75 per cent or less of maximum heart rate. Twenty of those 120 grams of glucose can come from glycerol (a by-product of fat metabolism) and the balance (easily) from gluconeogenesis in the liver. We don't have to rely on a single carbohydrate calorie to fuel our brains with glucose, or muscles with glycogen.

Unless you are an active growing youth, an extreme training athlete or a physical labourer depleting muscle glycogen stores daily, you probably don't ever need to consume more than 150 grams of dietary carbohydrate per day on average. Once you become fat adapted, you can probably thrive on far less. Many Primal Blueprint enthusiasts do very well on as few as 30–70 grams a day, even when engaged in an ambitious schedule of Primal workouts.

If you can limit carb intake to the bare minimum of what is necessary (or even up to 50 grams a day over that), and make up the difference with tasty fats and protein, you can literally reprogramme your genes to their evolutionary-based factory setting, the setting you had at birth. In only 21 days, you can transform yourself into an efficient, fat-burning organism, and can maintain that status as long as you send the right signals to your genes. The idea that you can become an efficient, fat-burning beast is the major premise of the Primal Blueprint eating and exercise strategies.

BECOMING A KETO-BURNING BEAST TOO!

We've already talked a little about the three major fuels that our bodies use for energy: carbohydrates (as glucose), fats (as free fatty acids) and protein (as amino acids). But now we're going to discuss a fourth fuel that our bodies evolved to metabolise when carbs are scarce: ketones or ketone bodies. Despite the undeserved misgivings some people have about ketones, it's one of the best energy management tools we humans have developed.

Ketones are an energy rich by-product of gluconeogenesis produced when the liver metabolises fat for gluconeogenesis. Ketones are typically considered by the sugar-burning world to be merely an emergency fuel, but they do much more. Your brain actually works more efficiently with ketones than with glucose, probably due to our ancestors' ability to access internal ketones rather than hard-to-find external carbs. When you become keto adapted, you will substantially decrease your glucose requirements over time. Your heart and other organs can also work very well on ketones and skeletal muscle can be trained to rely heavily on a mixture of fats and ketones as fuel for long periods, when you are exercising at 75 per cent of maximum heart rate or below.

Throughout any given day, the average person uses a combination of carbohydrates, fat, protein and ketone bodies as fuel. The relative amount of each fuel being used shifts in accordance with the signals you send your genes through diet and exercise choices. If you have programmed yourself to be a sugar burner up until now, most of your energy will come from glucose (the carbs from your meals or the glycogen stored in your muscles) and very little will come from fat (free fatty acids). Some amount of amino acids from meals or muscle tissue will contribute directly to fuel needs, while a bit more will be converted to glucose though gluconeogenesis. Finally, you'll derive a tiny energy contribution from ketones, so even a sugar burner is metabolising ketones on some level.

As you begin to shift away from a grain-based carbohydrate diet towards one that relies on healthy fats and proteins, you will send your genes new signals that up-regulate not only the systems that release fat from cells, but also those that improve fat-burning rates. You will get to the point where the vast majority of your daily energy requirements come from your new-found ability to tap into your stored body fat and to burn it easily. You'll also up-regulate the biochemical machinery that burns ketones more efficiently. We use terms like being 'fat adapted' and 'keto adapted' to describe these particular aspects of gene reprogramming. Adapting to your human 'factory setting' is truly existing in a 24/7 fat-burning zone.

You may find this scientifically interesting, but what are the real-life advantages of being fat and keto adapted? For one, you will no longer have an issue with storing excess body fat, and will easily arrive at and maintain your genetically ideal body composition. Second, you will experience less systemic inflammation, which is a disturbing by-product of being a sugar burner. Pumping too much glucose through your bloodstream every day compromises healthy cellular function, something that scientists and medical experts universally agree on, and use the catch-all term 'metabolic syndrome' to characterise. Finally, your body will adapt to survive for the rest of your life on fewer calories – without an increase in your hunger or a decrease in your daily energy levels. You will be less reliant and less hassled by the need to constantly find fuel to keep your brain and body running, and you will very likely extend your lifespan (and the quality of that lifespan), as this so-called 'calorific efficiency' is strongly correlated with longevity in every living organism.

The big problem when you are a sugar burner and you don't eat every few hours is that your blood glucose stores dwindle and you get tired and irritable. Your brain (which doesn't burn fat and isn't yet keto adapted) will frantically prompt you to become hungry and to seek out more carbs. If you don't refuel quickly, your brain will direct your adrenal glands to release hormones that stimulate the production of glucose in your liver. The fresh supply of glucose will help re-energise your brain and bloodstream, so you will feel better for a little while. However, because you are a sugar burner, you will break down lean muscle tissue in order to produce this glucose, and produce ketones as a by-product of this gluconeogenesis. Unfortunately, because you are not yet fat adapted or keto adapted, your brain and muscles haven't had the time to up-regulate the biochemistry necessary to effectively use these ketones for energy. Instead, they will leak out (the ketones – not your brains), unused, into your breath, sweat and urine.

This state of ketosis is not dangerous or even harmful (unless you are a type 1 diabetic, but that's a whole different biochemical process). It's just an indication that your body is producing ketones at a rate faster than it can use them as fuel. Maybe you've experienced this yourself or have noticed a sugar-burning friend who has skipped a few meals and now has that sweet 'acetone' breath. He or she is making ketones, but can't use them yet. On the other hand, when you become fat adapted and keto adapted, your body derives most of its energy from fats (either from a recent meal or from typical storage depots on your body), and your daily requirements for glucose drop dramatically. You can go for long periods of time without eating (either through a purposeful fast or by just forgetting to eat during a busy day), and you won't experience sugar burner side effects such as light-headedness, lethargy, bad moods and hunger. Hence, stress hormones aren't called into action, and muscle tissue is spared.

When you arrive at this blissful, genetically optimal state as a fat-burning and keto-burning beast your perspective about food and weight loss shifts forever. It feels a little like you have won a brand-new Prius in a raffle, because you have upgraded your body to a hybrid engine. You can either burn fuel from your tank (i.e. ingested calories) or use the internal battery (stored energy from body fat and ketones). You enjoy better fuel efficiency (fewer visits to the petrol station) and you will unload excess weight from your chassis in the process.

ENJOYING BESTIAL BENEFITS

If you've been a sugar burner for decades, you may struggle initially with deregulating your obsessive carbohydrate refuelling sessions. However, in a matter of a few weeks of sending the right signals to your genes through Primal eating, your appetite, hormone function, energy production, fat metabolism and other systems will self regulate. Worrying about calorie counting, portion sizes, meal timing, food combining, glyceamic load scores and other sugar burner madness will become irrelevant when you are a fat-burning beast.

After 21 days of gene reprogramming, you will be able to Intermittently Fast (IF) at any time, and achieve efficient weight loss/weight control as you become adapted to fat burning, gluconeogenesis, ketosis and inconsistent meal times. The secret is to moderate your insulin production by restricting processed carbs and sugars, and to obtain sufficient daily fats and protein calories from your diet to sustain lean muscle and general metabolic function. You will then be able to manufacture glucose or ketones as needed, and eliminate the fat storage pattern that occurs when you have a sugar-burning body.

In only a few weeks of Primal eating, your appetite, hormone function, energy production, fat metabolism and many other aspects of your day-to-day health will self regulate. Worrying about calories and portion sizes will become irrelevant when you are a fat-burning beast.

THERE IS NO REQUIREMENT FOR DIETARY CARBOHYDRATES IN HUMAN NUTRITION: SUMMARY

- Your 'factory setting' is to be an efficient fat-burning beast. You can manufacture glucose and ketones internally, rendering dietary carbs almost unnecessary.

- A high-carbohydrate diet interferes with these internal energy production mechanisms, making you reliant on dietary carbs (a 'sugar burner') until you can reprogramme your genes.

- Gluconeogenesis converts protein and fat into glucose in the liver. If you are fat adapted, it's an efficient energy mechanism (utilising dietary protein), but sugar burners can suffer from the breakdown of lean tissue during the process.

- You will need 150 grams per day of carbohydrate (except in extreme circumstances with mega-calorie burners), which can be obtained largely from vegetables, fruits, nuts and seeds. Grains, sugars and even pulses are unnecessary.

- Ketones are a by-product of using fat to fuel gluconeogenesis and can be an efficient source of energy, provided you are keto adapted. By contrast, sugar burners excrete valuable ketones in breath, sweat and urine.

- You can reprogramme your genes in 21 days of Primal eating, transforming yourself from a sugar burner into a fat-burning (and ketone-burning) beast.

SUCCESS STORY: TARA GRANT, TRAVIS AIR FORCE BASE, CALIFORNIA

At marksdailyapple.com, we have published hundreds of reader success stories dating back to 2006. Few stories are as successful as Tara's, whose frustrations, setbacks and her eventual healthy transformation provide valuable insights into the benefits of a Primal lifestyle.

Tara reports that she was healthy and active until the age of 24, at which point things started to fall apart: acne, allergies, boils, extreme weight gain, irritable bowel syndrome, itchy scalp, joint pain, reproductive problems and, not surprisingly, depression. She visited a number of doctors, who 'poked, prodded and bled' her, to no avail. 'The conclusion that all of the doctors shared was that I was over-reacting, hysterical and wasting their time. According to them, I was just fat and lazy – and probably a hypochondriac,' Tara relates.

Through the years, Tara's health problems continued to escalate. Tara finally took matters into her own hands, did research online and diagnosed herself with Premenstrual Syndrome (PMS), polycystic ovarian syndrome (PCOS), metabolic syndrome, endometriosis, a rare skin condition called hidradenitis suppurativa and depression. Doctors concurred and put her on Prozac, which only escalated her weight gain. Tara, standing at 175 cm (5 ft 9 in), topped out at 107 kg (16 st 11 lb) – eating a diet of 'nothing but simple, refined carbs ... I thought nothing of eating an entire plate of white rice for dinner, and nothing else.'

Tara experimented with a low-carb diet and quickly dropped 18 kg (40 lb). She embarked upon an intensive exercise programme – she 'practically killed' herself – to drop another 9 kg (20 lb) in time for her 2005 wedding. Alas, 'the minute I said, "I do", I gained five pounds.' For the next four years, Tara's life was a blur of hormonal problems, doctor visits and 'useless tests and pharmaceuticals'. In 2008, she had a 'miserable, sugar-soaked, bedridden pregnancy' during which she gained 36 kg (80 lb). Even five months post-partum, she was holding at 107 kg (16 st 11 lb), 'pissed off because I thought the weight was supposed to just melt right off' after pregnancy.

Tara's brother introduced her to the Primal Blueprint in May 2009. She discovered marksdailyapple.com community members who reported similar health symptoms and Primal-inspired health improvements. She eliminated grains and sugars and gradually drifted in a Primal direction, honouring the 80 per cent rule by 'planning treats once in a while and enjoying the hell out of them'. Meanwhile, Tara noticed that her hormonal imbalances were correcting – skin clearing up, reproductive cycles regulating and mood improving: 'I was incredibly happy all the time, and I was losing weight!'

With fat loss humming along at a slow and steady rate ('about a pound a week; I wasn't really trying that hard'), Tara's energy increased such that she could embark on a balanced exercise programme: CrossFit workouts a couple of times a week, some yoga, walking around the neighbourhood with her kids and even occasional sprints. 'I especially liked Mark's take on fitness – you don't have to do as much as the industry leads you to believe,' relates Tara.

Eighteen months into her Primal journey, Tara submitted her weight as 69 kg (lost 13 lb), 'less than I did in high school!' She says, 'Since my insulin and blood glucose are working properly now, going a whole day without food actually leaves me feeling fantastic and energised. I couldn't have imagined missing even a single meal in my old life! My depression, PMS, PCOS, endometriosis and everything else is just ... gone.'

Tara continues, 'My entire immediate family has now gone Primal and we have all had tremendous success. My mum and dad also now weigh less than they did in high school. My two-year-old twins are doing great on a (mostly!) Primal diet. They sleep through the night, are fairly mellow most of the time and are quite big for their age. Their first solid food was puréed chicken and apples. I'm currently working on becoming a certified personal trainer, so I can officially start helping others. In the meantime, I'm doing what I can through my blog and becoming involved in the Primal/paleo community. Mark Sisson, the Primal Blueprint and Mark's Daily Apple have changed my life, and I would like to pass it on to others.' Good luck Tara!

KEY CONCEPT 4:
80 PER CENT OF YOUR BODY COMPOSITION SUCCESS IS DETERMINED BY HOW YOU EAT

The right kinds of exercise are important for overall health and for developing functional strength and fitness, but I am certain that *at least 80 per cent of your body composition success is determined by how you eat*. Tens of thousands of Primal Blueprint users have confirmed this, and in a way, so have millions of devoted fitness enthusiasts who struggle and generally fail to reach an ideal body composition despite hours of weekly exercise and 'watching' what they eat. The problem is, they are eating grains, sugars and pulses. When you fully understand (we like to say, 'when you Grok') this key concept and eliminate the foods that promote systemic inflammation and fat storage and start eating Primally, you will automatically lose the bloating, because inflammation causes water retention and superfluous weight gain. This enables you to begin the process of burning off excess body fat until you eventually reach your ideal body composition.

Shedding 1 to 1.4 kg (2 to 3 lb) a week of steady fat loss is not unusual among Primal enthusiasts. Of course, results may vary according to your genetic predisposition to store fat, but when you moderate carbohydrate intake and insulin production to a genetically optimal level, you will lose excess body fat. This will happen no matter how predisposed your familial genes are to storing fat, as we saw with Paul and Chase Sheaffer on pages 26 and 64. Also, when you eliminate the foods which cause severe digestive problems and/ or autoimmune issues, you free up important nutrients that allow you to regain health and repair metabolic damage.

Unfortunately, we have been socialised to fixate on ineffective ways of controlling weight. We envy those with simple genetic good luck or with the exceptional discipline to perpetually restrict calorific intake and/or exercise like crazy. The rest of us seem resigned to a disturbing fate: Americans gain an average of 680 g (1.5 lb) and lose 227 g (½ lb) of muscle each year throughout their adult lives (ages 25–55). It may be hard to swallow the concept that the cereal, orange juice, wholemeal bread and pasta you eat have more influence on your waistline than your gym membership or the big numbers in your training log, but they do. Strong evidence – both scientific and anecdotal – points to the wildly excessive insulin production of the typical Western diet as a major variable in our collective failure to achieve body composition potential.

If you stand at the entrance to Disneyland or a Las Vegas casino, you will get a good cross section of the typical Western diet statistics: 68 per cent of American adults are overweight and 34 per cent obese. Stand at the starting line of a major marathon or observe an aerobics class at your local health club and things look a whole lot better, but you will still see a surprising range of body composition levels for such a driven group of exercisers. Many fit people carry 4.5, 6.8 or even 9 kg (10, 15 or 20 lb) of excess body fat despite training regimens of 10, 15 or even 20 hours per week! Such is the fate of the sugar burner. Consuming ample processed vegetable oils (rapeseed, safflower, sunflower, soya bean oils, etc.) in the name of pursuing better heart health certainly doesn't help their cause either. These unhealthy fats send signals to their genes that promote inflammation, insulin resistance and other metabolic damage.

Sure, there are some exceptions, like the driven and genetically gifted types who can train long hours, refuel on carbs and not add much body fat (hey, I was one). Unless you love to work out incessantly and have exceptional familial genes, however, a sugar-burning existence is unsustainable and ridiculous. You are running on a literal and figurative treadmill of insidious weight gain over the years, with little to no understanding of the true cause. Like an experimental lab rat, you are behaving as directed by Conventional Wisdom and destroying your health in the process.

Nutrition and medical experts now believe that a vigorous exercise programme, while offering assorted health and fitness benefits (provided you don't overdo it), stimulates increases in appetite and calorie consumption that result in a draw when it comes to weight management (we will look at this in detail in Key Concept 7). Think about it: you assume that carbohydrates are the basis of a healthy diet, so you eat lots of them. As a result, your insulin levels rise and store the excess carb calories as fat. So you think to yourself, 'I have to exercise more to burn off this fat,' but because you are a sugar burner, your muscles prefer to burn glucose and glycogen instead of fat. This depletes your glycogen stores every time you exercise, making your brain think, 'If this fool is going to try this again tomorrow, I'd better stock up on carbs and refill my glycogen tonight.' Hey presto, you eat even more carbs over the next 24 hours, just so you can exercise hard again the next day! And the calorie-burning, carb-eating, fat-storing cycle continues.

> *Exercise stimulates an increase in appetite and calorie consumption such that it results in a draw when it comes to weight management.*

Your workout routine and lifestyle obviously have some influence on your body composition potential – just not as much as most people assume. I estimate a sensible Primal exercise program contributes about 10 per cent of influence on your ultimate genetic potential, while an extreme training regimen (heavy CrossFit, competitive endurance or team sport athlete) contributes another 5 per cent. Complimentary lifestyle factors – particularly getting adequate sleep to regulate appetite hormones – contribute another 5 per cent. Fully 80 per cent of your body composition is determined by how you manipulate hormones and gene expression through the foods you eat. Owning this fact can be one of the more empowering aspects of The 21-day Total Body Transformation.

IT'S MOSTLY ABOUT INSULIN

When we look closely at how we can manipulate hormones and genes to arrive at our ideal body composition, we see that insulin has a profound influence. Studies have investigated the influence of various other hormones (glucagon, leptin, ghrelin, PPY, T3, adrenaline, epinephrine, cortisol, testosterone, human growth hormone and many more) on hunger, satiety, appetite, inflammation, obesity, diabetes and general metabolism, but the most important thing for you to know is that insulin sits at the very top of this hierarchy.

Insulin exists in all living organisms and is one of the oldest hormones in nature. We've talked about insulin's role in fat storage, but insulin also plays a role in thyroid function, stress hormone production, sleep cycles, appetite regulation and the metabolisation of cholesterol and triglycerides (the storage form of fat). When you are able to optimise insulin production, sex hormones are delivered properly to target organs, cholesterol assists with energy and hormone production, and appetite, sleep and thyroid hormones become balanced. Insulin is considered the 'master hormone' for its critical and comprehensive role in transporting nutrients from the blood-stream into the cells in your body, but some of its most important action takes place in muscle cells and fat cells, which is where you see the most obvious effects.

It's an undisputed scientific observation that within any species, individuals who secrete the most insulin over a lifetime live the shortest lives. That fact alone should be enough to prompt you to reprogramme your genes to operate more efficiently at lower insulin levels. Our ancestors' pre-civilised diet of meat, fish, eggs, insects and the occasional wild vegetable, fruit or root yielded minimal insulin production over a their lifetimes. Even fruits, the one major source of dietary carbs during Grok's time, were generally eaten only during narrow ripening seasons, and were far more fibrous and less sugary than today's highly cultivated, sweetened varieties, prompting very little insulin secretion.

This predominantly fat-burning existence programmed our human genetic recipe to favour *insulin sensitivity*. This means that receptor sites in your cell membranes can easily use small amounts of insulin as a key to unlock special-access pores, enabling the assimilation of desired nutrients into the cell: fatty acids, glucose, amino acids, vitamin C and various other molecules. In healthy, fit people, muscle cells are the largest benefactor of insulin's actions. Amino acids are granted access to repair and increase muscle size and strength, and glucose is allowed inside for immediate energy or conversion into glycogen for storage.

When you chronically overproduce insulin as a sugar burner, and fail to routinely empty and restock muscle glycogen through adequate exercise, your muscle cells become *insulin resistant*. Insulin resistance occurs when muscle and liver cells become desensitised to insulin's storage signals due to excessive production. Instead of unlocking pores, those cells put out a 'No Vacancy' sign, because they are typically already full of glycogen.

The repercussions of insulin resistance are severe and wide ranging. Ironically, when the insulin is no longer effective at normal levels, your pancreas thinks more will do the trick and pumps out an even higher volume. High insulin levels wreak havoc in many ways. Cholesterol and triglycerides become oxidised and inflamed, initiating the process of atherosclerosis, or hardening of the arteries. Appetite hormones get thrown out of balance, causing you to continue eating even when you should feel satiated. Fat cells stay sensitive well after muscle cells have become resistant, so the excess sugar and other calories wind up as stored body fat. Melatonin/serotonin cycles get messed up, making you feel groggy and grumpy in the morning, and craving sugar in the evenings. Becoming insulin resistant can directly lead to metabolic syndrome and type 2 diabetes. You just don't live as long, or as well, thanks to sugars, grains, pulses and the massive lifelong insulin overdose caused by the typical Western diet.

> *Insulin resistance occurs when muscle and liver cells become desensitised to insulin's storage signals due to excessive production. This condition promotes obesity, accelerated ageing, atherosclerosis, sleep deprivation and type 2 diabetes.*

Insulin resistance is strongly linked with obesity, but even those without excess body fat suffer negative consequences from a high insulin-producing diet. Those seemingly fortunate people who appear skinny don't tend to store much visible fat under the skin, but rather in deeper tissues and organs (known as visceral fat), which brings an even greater risk of disease. Furthermore, with fat storage genes switched off, blood glucose levels can spike, increasing disease risk. This excess blood sugar, with nowhere to go, binds with proteins that cause all manner of nerve damage, arterial damage, vision damage and much more. Besides tracking your disease risk factors through the Body Mass Index (BMI) charts, it's advisable to test your blood for triglycerides, fasting blood glucose and insulin, C-reactive protein and small, dense LDL, a special type of cholesterol molecule that promotes atherosclerosis. If these markers are out of normal ranges, it can foretell risks beyond the numbers on the scale or the statistics produced by a routine physical examination.

Western medicine is adept at identifying the particular health problems caused by insulin overproduction, but errs in applying a targeted pharmaceutical approach to a big picture problem. Statins, sleep medication, thyroid medication, anti-ageing hormones and other powerful prescription drugs are commonly dispensed to counteract imbalances that originate with excess insulin production. These drugs artificially manipulate gene expression as intended, but do nothing to address the root cause of typical Western diet-related health problems. Furthermore, prescription drugs often exacerbate these problems by destabilising natural hormone production to the extent that the user becomes reliant on medication to achieve repeated temporary relief at the expense of long-term health.

The good news is that your day to day health is greatly influenced by the quality of your last meal. Eat just a few low insulin-producing meals, and you will likely experience an improvement in appetite, energy levels, sleep cycles and other sensations of vitality. Establish a pattern of insulin moderation for 21 days, and several pounds of excess body fat might just melt away without the typical deprivation and restriction associated with fat reduction efforts. Over the long term, moderating insulin production could save your life – neutralising the pathologies of metabolic syndrome, systemic inflammation, glycation (excess glucose damaging protein molecules, leading to assorted disease and dysfunction), obesity and heart disease that originate in large part from being a sugar burner.

> *Your day to day health is greatly influenced by the quality of your last meal. Eat just a few low insulin-producing meals, and you will likely experience an improvement in appetite, energy levels, sleep cycles and other sensations of vitality.*

THE EXHAUSTION EPIDEMIC

When you chronically overproduce insulin as a sugar burner, you overstress various delicate hormonal and metabolic mechanisms that have been hard wired through evolution to thrive in the fat-burning zone. A vicious circle leading to exhaustion transpires as follows: first, the ingestion of processed carbs elevates blood glucose levels, creating an immediate boost in energy, mood and cognitive function. Within minutes, your elevated blood glucose prompts the pancreas to release insulin into the bloodstream. This is an essential function, because excess glucose in the bloodstream is toxic and can quickly become life threatening if not removed (as experienced by diabetics). When insulin does its job and removes glucose from the bloodstream, this 'sugar crash' suddenly makes you feel sluggish, moody and unfocused.

> *The combination of a blood glucose spike, insulin-triggered glucose crash, stress response pumping more glucose into the bloodstream temporarily and eventual stress hormone crash leads to a physical, mental and emotional lull best described as 'burnout'.*

This sudden drop in energy is perceived as a stressful event by your body, triggering the familiar fight or flight response. When the fight or flight hormone cortisol floods your bloodstream, muscle tissue is converted into glucose through gluconeogenesis, giving you the quick energy you crave, but often causing you to feel jittery and hyper. After some time, the mood- and energy-elevating effects of cortisol wears off. After all, the fight or flight response is designed to produce brief bursts of high energy and heightened function for emergency situations, not as a recurring component of your daily metabolic rhythm.

The combination of a blood glucose spike, insulin-triggered glucose crash, temporary glucose pump and eventual stress hormone crash leads to a physical, mental and emotional exhaustion often described as 'burnout'. Over time, the daily roller coaster of too much glucose and insulin in the bloodstream promotes systemic inflammation, setting the stage for assorted health problems and diseases.

SUCCESS STORY: PAUL (AND CHASE) SHEAFFER, PENSACOLA, FLORIDA

Paul, a 26-year-old software developer two years into his Primal journey, sets the opening scene for his story as follows: 'As I went through college, my entire life consisted of sitting in front of computers, playing TV video games and watching movies. I felt like a slob and I lived like a slob. While I was generally a happy person, I always felt like I was doing myself a disservice. I wanted to become more involved with people, especially the opposite sex, and I wanted people to start taking me more seriously. Unfortunately, a mixture of World of Warcraft (mass participation online game), a steady diet of cheese puffs and caffeinated fizzy drinks and horrible social anxiety caused by my weight and general appearance made this very difficult to do.'

By the time Paul decided to make some changes in his life, he was hovering around 136 kg (21 st 6 lb). A most unlikely motivator came in the form of a supermarket potato salad container. 'I ate the entire container, endured three days of horrible sickness from food poisoning, and came out 15 pounds lighter. At one point I passed out, face down on a tile floor and shattering my glasses. When I saw myself in the mirror without the glasses (inspired by a slightly thinner waistline from the ordeal), I made up my mind right there that I would fix myself and become a member of society again.'

Paul continues, 'I immediately jumped on to the wholegrain bandwagon and started eating oatmeal by the boat load, whole grain bagels every morning, fibre crackers with lunch, and every other classic Conventional Wisdom food you could imagine. A few months into this lifestyle, I developed a crippling case of diverticulitis. I had a foot of my colon surgically removed at the age of 24.'

Paul recovered and continued to eat a grain-based diet, but fortunately, he encountered some positive influences by joining a local CrossFit fitness facility. 'I was completely humbled and embarrassed to see myself next to the hard bodies, but I used them as inspiration to keep pushing myself harder.' He also heard the fittest members extolling the Primal eating style, and decided to start eating that way. After only two months, Paul shed a remarkable 23 kg (50 lb). Two more months and he'd lost 39 kg (85 lb), and now weighed 98 kg (15 st 5 lb). 'Everyone around me was amazed. I was feeling the best I ever had in my life ... girls no longer looked the other way when I walked into a room!'

In 15 months, Paul went from 136 kg (21 st 6 lb) to a chiselled 68 kg (10 st 10 lb) – half the man he once was, but with double the confidence. 'Everyone around me was amazed. I was feeling the best I ever felt in my life!'

Over the course of about 15 months, Paul created a chiselled 150-pound physique – half the man he once was, but with double the confidence. 'I was able to begin making friends, get over my social hang-ups, and actually go out and do things in public, in the fresh air – even compete in martial arts – and not be ashamed. I have transformed myself into a powerful, lean, and self-satisfied person. I now train in Brazilian Jiu-Jitsu and have more friends than I've ever had in my life. If it wasn't for the *Primal Blueprint* book and marksdailyapple.com – with its amazing forums – to keep me on track and answer my questions, I do not believe I would have ever reached where I am now, especially as quickly as I did.'

Paul magnified the impact of his success by inspiring his identical twin Chase to go Primal. In addition to Chase's rapid fat loss of 70 pounds in five months that we previously mentioned, he has corrected medical conditions (tachycardia and high blood pressure) to the extent that he can eliminate prescription drug use and self-regulate heart beat and blood pressure. Chase believes that 'this change has been 100 per cent due to diet. My medical issues started before I went Primal and decreased rapidly when I made the shift.'

Paul continues, 'Being an identical twin is a one-of-a-kind relationship. No matter what we do independently for our entire lives, we always have this huge genetic connection. I definitely thought about Chase throughout my Primal transformation, and nudged him whenever I could without pressuring him or being judgemental. There is no better way to help others than to lead by example – particularly when you happen to share the exact same genetic code!' Today Paul and Chase earn a handsome living as impersonators and stunt doubles for Hollywood star Matt Damon ... just kidding! Maybe ...

DIALLING-IN RESULTS WITH THE CARBOHYDRATE CURVE

The impact of various levels of carbohydrate intake, and consequently insulin production, are shown on the Primal Blueprint Carbohydrate Curve below. When you are at your ideal body composition, eating Primally will default you into the Effortless Weight Maintenance Zone. When you are interested in losing weight, you can make a more focused effort to eat in the Sweet Spot Zone. This will enable you to reduce body fat at a rate of 1.8 to 3.6 kg (4 to 8 lb) per month (rate depending on sex, weight and fitness level) until you reach your ideal body composition. On the other hand, eating the typical Western diet will default you into the Insidious Weight Gain Zone or Danger Zone.

The 50-gram/200-calorie range within each zone on the curve allows for individual disparities in body weight and metabolic rate. A smaller woman might align with the lower end of the range, while a larger man might find the upper end of the range more accurate. While the curve offers general recommendations, tens of thousands of marksdailyapple.com and *Primal Blueprint* readers have proven that the curve is quite accurate. The Primal eating strategy emphasises an intuitive and sporadic approach to eating. The values on the Carbohydrate Curve are averages – you only need to concern yourself with landing in the desired zone on the curve *on average* for a week's time or a month's time. The same is true for meeting your average daily protein requirement to preserve or build lean muscle tissue. Relax, enjoy your meals, honour the general Primal Blueprint philosophy, and do not by any means obsess over your macronutrient values at each meal!

Note for athletes: if you are involved in a devoted exercise regimen, you can adjust your curve values upwards to fuel particularly hard workouts. Depending upon your individual variables and body composition goals, you can experiment with adding up to 100 grams of daily carbohydrates for each hour of *vigorous* exercise (over 75 per cent of your max heart rate) that you engage in. Experiment is the operative word. If you are struggling to reach body composition goals despite a devoted exercise programme and Primal eating efforts, you may have to track the association between carb intake and body composition, as described in the Primal Leap weight-loss programme at primalblueprint.com.

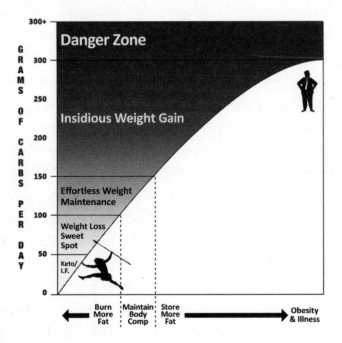

0 to 50 grams per day – Ketosis/IF Zone: Acceptable for occasional one to two day IF efforts toward aggressive fat loss (or longer term, for medically supervised weight-loss programmes for the obese and/ or type 2 diabetics), provided adequate protein, fat and supplements are consumed. Eating in this zone is not recommended as a long term practice for most people due to the resultant deprivation of high nutrient value vegetables and fruits.

50 to 100 grams per day – Weight Loss Sweet Spot: Minimises insulin production and accelerates fat metabolism. By meeting average daily protein requirements (0.7 to 1 gram per pound of lean mass – range depends on activity level), eating nutritious vegetables and fruits, and staying satisfied with delicious high-fat foods (meat, fish, eggs, nuts, seeds), you can lose about ½ to 1 kg (1 to 2 lb) of body fat per week in the Sweet Spot.

100 to 150 grams per day – Maintenance Zone: Allows for genetically optimal fat burning, muscle development and weight maintenance. Rationale is supported by humans eating and evolving in this range (or below) for two million years. Enjoy carbohydrates from abundant servings of vegetables, strategic consumption of fruit and incidental carbs from nuts, seeds, high-fat dairy products, and occasional dark chocolate.

150 to 300 grams per day – Insidious Weight Gain Zone: Continuous insulin-stimulating effects prevent efficient fat metabolism and contribute to widespread health conditions. The de facto recommendation of many popular diets and health authorities – including the United States Department of Agriculture (USDA) food guide pyramids! – is in this zone, despite clear indications that this promotes the development of metabolic syndrome.

Some chronic exercisers and active growing adolescents may find that they can eat at this level for an extended period without gaining fat, but eventually fat storage and/or metabolic problems are highly probable. This insidious zone is easy to drift into, even for health-conscious eaters, when grains and pulses are a dietary centrepiece, obligatory vegetables and fruits are enjoyed and sweetened beverages and treats leak into the picture here and there. Despite well-intentioned efforts to moderate intake of fat, sweets and total calories, many people still gain about ½ to 1 kg (1 to 2 lb) of fat per year for decades as a result of carb intake in this insidious range.

300 or more grams per day – Danger Zone: This is where the average American falls, thanks to following the official USDA dietary guidelines (which suggest you eat 45 per cent to 65 per cent of daily calories from carbs), and routinely adding sweetened beverages, packaged snacks and desserts into the picture. Extended time in this zone results in almost certain weight gain, metabolic syndrome, obesity, type 2 diabetes and other widespread health problems. For people in this zone, immediate and dramatic reduction of sugars, grains and pulses is critical.

IDEAL BODY COMPOSITION RESULTS MAY VARY ... FROM GREAT TO AMAZING!

When we talk about ideal body composition in the Primal Blueprint, we are talking about *your genetic ideal*, not your dream of gracing the cover of *Men's Health* or *Men's Fitness* magazine at single-digit body fat. As we mentioned before, your parents (and your parents' parents) contributed a range of possible outcomes to your human recipe. You may not be able to achieve 13 per cent body fat as a woman, or 8 per cent as a man, without an unrealistic and unpleasant amount of pain, suffering and sacrifice. But that's okay. Let's say you make great progress with the Primal Blueprint and lose ½ kg (1 lb) or so each week – more or less effortlessly –

for weeks or months until you hit a plateau, perhaps 21 per cent body fat for a woman, or 15 per cent for a man. That doesn't mean that you can't drop down even further over time, but for now, you might be at your genetically ideal body composition. Here, your familial genetic recipe is telling you, 'I like this weight. I feel great, I never get sick any more, I'm stronger, I handle stress easier, my clothes fit better, I have all the energy I need, I can maintain this weight easily without "dieting", and I'll live a long time at this weight.' What's wrong with that?

We say at marksdailyapple.com that in order to have what you want, you have to want what you have. When it comes to familial genes, it's important to love yourself for who you really are, and to recognise that you may have ultimate limitations compared to other genetic specimens. By living and eating Primally, you can achieve the highest potential of your personal genetic recipe. When you acknowledge this, you are free to enjoy life and food without guilt or disappointment. From there, you can decide if it's truly worth the time, energy and effort to get even leaner and stronger. As many people discover, the increase in sacrifice to add those extra 4.5 kg (10 lbs) of muscle or to drop more body fat just isn't worth it.

Fixating on the genetic freaks gracing magazine covers or playing on professional sports teams is not recommended, because these 'role models' generally engage in extreme diet and exercise regimens that are undesirable and unhealthy to emulate. Instead, hark back to the age when you looked and felt your best. Perhaps you were an active youth or a school athlete? If not, consider a fit sibling as a reference point. Depending on your starting point, you can achieve your personal genetic potential quickly: 45, 90 or 365 days from now.

Take a moment to reflect on the remarkable stories of Paul Sheaffer (see page 64), Tara Grant (see page 54), and Timothy Williams (see page 84). They, and hundreds of other success stories at marksdailyapple.com, are real people leading busy, hectic lives, possessing a range of genetic attributes, and transforming their bodies by living Primally. Results may indeed vary based on genetic factors, but this caveat is mostly relevant to what you can control: you'll have a range of bad luck to terrible luck when you mismanage your genes. When you eat, exercise and live Primally, you'll look and feel your very best – it's that simple!

80 PER CENT OF YOUR BODY COMPOSITION SUCCESS IS DETERMINED BY DIET: SUMMARY

- Body composition is 80 per cent dependent on diet. Excess body fat reflects familial genetic predisposition *combined with* dietary insulin production.

- Insulin is the 'master hormone', responsible for transporting nutrients and hormones through the bloodstream to target cells and organs.

- Excess insulin production from a grain-based typical Western diet inhibits fat metabolism, disrupts hormone and immune function, accelerates ageing, and promotes systemic inflammation and disease. Moderating insulin production promotes efficient fat metabolism, optimal hormone and immune function and longevity.

- The Primal Blueprint Carbohydrate Curve reveals that eating 150 grams per day or less results in effortless weight loss or maintenance; eating over 150 grams (per typical Western diet recommendations) results in life-long insidious typical Western diet gain and accelerated disease risk.

- Your genes compensate for a high insulin-producing diet by triggering the fight or flight response, resulting in a 'burnout' cycle of glucose spike-insulin crash-stress response, and finally exhaustion. Alas, skinny people suffer too from burnout, accelerated ageing, and disease pathologies.

KEY CONCEPT 5:
GRAINS ARE TOTALLY UNNECESSARY
(AND SO ARE PULSES, FOR THAT MATTER)

There is simply no good reason for you ever to eat grains again, and plenty of reasons to completely eliminate all forms of grain foods from your diet. In fact, as I emphasise in my seminars, the *only* reason to make grains a part of your eating strategy is as a cheap source of calories that easily converts to glucose. Grain foods (wheat, rice, corn, pasta, bread, cereal, muffins, crisps, baked goods, etc.) elicit a high insulin response, have far less nutritional value than Primal foods and contain anti-nutrients (basically toxins) that promote inflammation and compromise digestion and immune function. Grains are merely a convenient source of calories that are easy to harvest, store and process into all manner of high-profit packaged, baked and frozen goods.

While the major focus here is on grains since they are the foundation of the typical Western diet, I must also mention pulses in the same breath. Beans, lentils, peanuts, peas and soya products are slightly less offensive than grains, but they also contain objectionable anti-nutrients and stimulate excessive insulin production. Pulses are yet another cheap source of modern calories that are simply unnecessary – and quite possibly counterproductive – to health, energy and weight management.

Granted, not eating grains, sugars and pulses may cause some initial discomfort and may also require tremendous focus and discipline to maintain over the long term. As you commence your transition away from a grain-based diet, you can make things easier by surrounding yourself with nutritious Primal foods whenever you need a snack or hanker to return to your old sugar-burning ways. Even though eliminating processed carbs may leave a void in your daily calorific intake, not to mention your kitchen, there is no reason you should ever feel deprived or depleted when you can replace them with abundant servings of delicious and satisfying Primal foods.

It's necessary to fully commit to this dietary transition in order to normalise your insulin production and switch to the fat-burning beast you can be. Half-hearted efforts to restrict grains, sugars and pulses will compromise the goal of reducing systemic inflammation, and healing digestive or metabolic damage. I must admit I've encountered plenty of unfavourable responses on this issue, with people arguing that they simply can't imagine life without various grain staples in their diet. The influence of comforting food rituals, warm bready smells and traditions is not to be discounted. I certainly have fond memories of stacks of blueberry pancakes at Sunday brunch, best enjoyed after a 32-km (20-mile) run in days gone by. But consider what you are giving up – is it really the taste of grains that you can't live without? A steaming hot bowl of plain porridge, plain pasta or brown rice? These are hardly foods that anyone would consider rich and satisfying by themselves. It's the *stuff you put on grains* that make them taste good, and you can easily recalibrate your notion of comfort foods and meal traditions when you go Primal.

Can't live without the bland taste of grains (plain oatmeal, pasta or brown rice)? It's the stuff you put on grains that make them taste good, and you can easily recalibrate your notion of comfort foods and meal traditions when you go Primal.

SUGAR-COATING GRAINS
(AND PULSES FOR THAT MATTER)

I don't spend much time discussing sugar per se in this book, as the drawbacks of consuming junk calories are well understood, and it's assumed that sweet foods and drinks should be virtually absent from your diet if you want to be healthy. The ingestion of sugary foods and beverages is the ultimate affront to your genes – it promotes systemic inflammation, causes an instant suppression of immune function and instigates the sugar high-insulin crash-stress hormone spike-burnout cycle discussed previously (see page 64). What I want you to get is that grains and pulses are not much different from straight sugar to your body. In fact, nearly every form of carbohydrate that you consume eventually winds up in your bloodstream as glucose. Your brain and your muscles cells can't tell whether the latest dose of glucose is from a bowl of porridge oats or a bowl of sugar! It's all just glucose to your body.

You may be familiar with the concepts of glycaemic index and glycaemic load, balancing macronutrients at meals into target 'zones', pairing quick-burning foods with slower-burning foods, snacking frequently with small portion sizes and other such nonsense we've already sufficiently criticised. All of this sugar burner folly still compromises optimal gene expression, which is only possible when you become fat and keto adapted.

Whether you start your day with a 380-calorie breakfast of Cheerios®, skimmed milk, banana and orange juice, lunch on a 380-calorie serving of brown rice and split pea soup, or stop at a snack bar to wolf down 380 calories of milk shake and a half-bag of Skittles®, your body will ultimately have to produce a requisite amount of insulin to deal with every gram of glucose generated by the ingested carbohydrate. A steady insulin drip to deal with the more complex 'slow-burning' carbohydrate grain foods is widely viewed as preferable to the quick spike junk food option, but we must rethink even this seemingly obvious conclusion. With the steady insulin drip, you apply chronic stress to your metabolic and endocrine systems, and inhibit fat metabolism over a longer period than what might have transpired with the Skittles and milk shake. And while the typical Western diet breakfast and lunch meals deliver more vitamins and other micronutrients than the snack bar offering, the scale of this advantage is inconsequential – dwarfed by the advantages of eating nutrient-dense plant and animal foods. Essentially, we are debating the lesser of three evils – with my sincere apologies to the cereal, dairy, rice and milk shake lobbies!

Your genes are adept at handling brief, intermittent stressors – a plunge in cold water, running a few all-out sprints, an occasional late night of revelry, an airplane trip across time zones or even an occasional slamming down of a milk shake. A junk food binge elicits a quick spike of insulin, followed by an energy lull, and then – if you are fat adapted from a general pattern of Primal eating your blood soon regulates to an optimal balance of insulin and glucose. In contrast, the chronic day-in, day-out stress of excess glucose and insulin in the bloodstream from typical Western diet eating leads to adrenal burnout and systemic inflammation.

The *Primal Blueprint Cookbook, Primal Blueprint Quick and Easy Meals,* marksdailyapple.com and various other cookbooks and Internet resources honouring the Primal/paleo/low-carb eating movement offer delicious suggestions, including clever substitutes for popular grain-based recipes. Spaghetti and meatballs? Keep the delicious meatballs and sauce, and pan-fry some thin slices of squash, carrot and courgette as a colourful substitute for pasta. Enjoy your BLT for lunch but hold the bread and make it a lettuce wrap – a LBLT – 'label it'! You can explore an exciting new world of culinary options without feeling like you have to sacrifice any gustatory pleasure.

WHOLE GRAINS AREN'T A WHOLE LOT BETTER ... AND MIGHT EVEN BE WORSE!

There seems to be universal agreement that refined grains such as sugary snacks and treats, white flour products and sweetened beverages are unhealthy. Even the food industry's sneaky inclusion of trans and partially hydrogenated fats and high-fructose corn syrup (HFCS) into all kinds of processed foods and drinks over the past few decades is being moderated thanks to consumer outcry. Fizzy drink machines are getting banned from schools and headline stories are connecting excess sugar consumption with obesity, attention deficit hyperactivity disorder (ADHD), allergies and other maladies. Well, that's a start.

Always trying to do the right thing in the name of health and disease prevention, Conventional Wisdom eagerly recommends you switch from refined grains to whole grains. Unfortunately, even a whole grain-based diet cannot save you from the perils of being a sugar burner, as we covered in the sugar-coating grains on page 75. Furthermore, whole grains contain anti-nutrients in the form of lectins, glutens and phytates that can compromise health in other ways. Yes, you need to give up your whole grains too.

Whole grains are comprised of three components: bran (fibre), germ (oil) and endosperm (starch). Refined grains have been stripped of their bran and germ to leave only the starchy endosperm. While whole grains have more vitamins, minerals and fibre than refined grains (this 'more' is inconsequential in comparison to nutrient-rich Primal foods), they also contain higher levels of anti-nutrients than refined grains. Furthermore, experts such as Konstantin Monastyrsky, author of *The Fibre Menace*, believe that a grain-based diet results in excessive fibre intake, resulting in nutrient depletion and digestive irregularities – the exact opposite of what we had been led to

believe. If getting enough fibre is a health concern for you, be confident knowing that eating reasonable amounts of vegetables, fruits,
nuts and seeds provides plenty of fibre to ensure digestive health, as
it has for two million years. And be open to the idea that your efforts
to obtain more fibre (via over-the-counter psyllium drinks and grain-
based eating) could actually be making you *irregular*.

Those who suffer from wheat allergies or gluten intolerance
understand how a seemingly innocuous slice of bread or plate of
pasta can throw your whole life out of balance. I'd argue that *all of
us are intolerant to the anti-nutrients in whole grains* at some level,
even if we don't experience acute symptoms. Lectins are natural
plant toxins that suppress immune function, interfere with normal
protective gut barriers and promote inflammation by allowing undi-
gested protein molecules to infiltrate your digestive tract and trigger
an autoimmune response – a situation characterised by the familiar
term, 'leaky gut syndrome'. You may not experience any major acute
symptoms, but long-term damage is being done to your delicate
digestive tract by these toxic foreign agents.

> *All of us are intolerant to the anti-nutrients in whole
> grains at some level. Even if you don't experience any
> major acute symptoms, a grain-based diet will tend to
> suppress immune function, compromise digestive health
> ('leaky gut'), and promote systemic inflammation
> (e.g. '-itis' conditions).*

Gluten, a type of lectin, triggers a pro-inflammatory condition in the
body, disturbing healthy immune function and promoting all manner
of inflammation-related health problems (skin, joint, reproductive,

allergies, assorted '-itis' issues), and other serious diseases over time. Coeliac disease is the most severe manifestation of lectin and gluten intolerance, and affects millions of Americans. Some milder forms of gluten intolerance are still believed to affect an additional third of all Americans, yet a significant portion of coeliac and gluten intolerance sufferers may not even realise their afflictions. Of course, gluten is most prevalent in wheat, and wheat is the most prevalent grain in the typical Western diet.

Phytates are indigestible agents that bind with and inhibit the absorption of important nutrients in the digestive tract, such as calcium, magnesium, iron and zinc. Phytates are found not only in whole grains, but in pulses, nuts and seeds. Eating small amounts of phytates is not problematic, since Primal eating delivers plenty of minerals. Excessive grain intake, however, as often seen in vegetarians and vegans, can promote mineral deficiencies due to the high levels of phytates in these foods. Millions of women take supplemental calcium to prevent osteoporosis, but also consume lots of 'healthy' whole grains, and then wonder why their bone density doesn't improve. It could be the high levels of phytates from grains inhibiting their absorption of calcium.

Those fortunate enough not to experience severe intolerance to anti-nutrients still suffer at a sub-clinical level such that digestive issues (bloating, bowel irregularities such as irritable bowel syndrome, colitis, constipation, gas), mild to nagging inflammatory conditions throughout the body and frequent immune system disturbances are seen as normal. Many Primal enthusiasts have experienced (as related in compelling Success Stories published here and on marksdailyapple.com) rapid and immediate improvements in overall health to the extent that lifelong health issues requiring prescription medication are reversed in a matter of weeks. Often, an individual's sense of general health becomes recalibrated so that 'exceptional' becomes the new normal!

WHEN IT COMES TO PULSES, DON'T ASSUME ...

Pulses include beans (black, kidney, pinto, soya and derivative products such as tofu, etc.), lentils, peanut and derivative products such as peanut butter, and peas. Pulses contain many of the same undesirable lectins as grains. Furthermore, pulses are generally not the great sources of protein they are sometimes reputed to be, but, like grains, are abundant sources of cheap carbohydrates that easily convert to glucose. Ironically, one of the main features of pulses touted by some dietitians is the idea that they contain lots of fibre and may assist in bowel health. However, once again, the fibre in beans and most other pulses is unnecessary and possibly counterproductive when you can satisfy your need for fibre with adequate vegetables. And really, who needs the gas?

Pulses are not something we evolved to eat. Not only do pulses have to be soaked, sprouted, denatured and cooked to be eaten at all, in many cases they are poisonous if they are not extensively pre-treated. In fact, one of the most deadly poisons known to man is ricin. An extract of castor bean, it has been used in military attacks for decades. Soya, which has become one of the largest crops in the world, and is used in countless processed foods, is not only a source of lectins, but has been also shown to have phyto-oestrogenic properties (effects that mimic oestrogens in men and women, and may disrupt normal sex hormone cycles). In the opinion of many experts, soya, and everything made from it, is one of the most offensive foods in the typical Western diet.

FOOD FOR THOUGHT

The massive global population expansion in the last few generations – for better or for worse – has been enabled by grain-based diets. Today, half the world obtains half of their calories from bread! Citizens of these impoverished nations suffer from widespread deficiencies of even the most basic nutrients. Furthermore, helter-skelter population growth compromises overall quality of life and the planet's natural resources, creating a complex public health and humanitarian dilemma. Those of us fortunate enough to live in relative comfort with abundant dietary options are compelled to reflect upon whether we are properly honouring these privileges.

While starving Third World citizens can be excused for eating whatever calories they can get their hands on, eating a diet of 71 per cent processed foods is an abomination for inhabitants of the wealthiest nations on earth. Our continued consumption of these products enables multinational food conglomerates to escalate marketing and manufacturing efforts and further entrench processed foods into modern culture.

I have deep respect and appreciation for the green ideals of sustaining oneself with minimal drain on the planet's natural resources, and believe that eating Primally is the best way to achieve global health and well-being. Vegans will take me to task for promoting the consumption of any kind of animal products, and I agree that there are moral, sustainability, environmental and physical health objections to consuming intensive factory farming products. However, I also advocate a pragmatic approach of doing the best you can under the existing circumstances of availability, budget and preference. I make every effort to emphasise organic, grass-fed animals over intensive factory farmed animals, but I would definitely eat a factory farmed animal product before I eat a slice of bread or a scoop of brown rice. In Action Item 3 we will get deeper into this issue as we detail the spectrum of best to worst choices in every food category.

If you are resistant to getting rid of pulses and grains from your diet, let's admit that pulses are slightly less offensive than grains, as they offer a bit more nutritional value and fewer anti-nutrients. However, in the context of correcting disastrous typical Western diet eating habits and insulin overdoses, pulses are simply not necessary – and can compromise your health and weight-loss efforts.

If you have any excess body fat, suffer from any sort of digestive irregularities or inflammatory conditions, it's worth testing the effects of a total elimination of grains, sugars and pulses from your diet for at least 21 days. See how much better you feel. Then, if you're at a party and see a plate of fresh veggies with a bowl of hummus (made from chickpeas) and are inclined to take a few dips, you certainly don't need to lose any sleep over it. Just remember that all of your dietary choices exist on a continuum, with the ultimate goal of aligning as closely as possible with optimal gene expression. When I see a grain or pulse option that might be of interest, I weigh it against having one more delicious chunk of steak, and the decision is no contest.

TRY A 21-DAY GRAIN-AND-PULSE-FREE TEST

I was one of those people who thought I could eat grains for ever and without penalty. When I completely eliminated grains from my diet back in 2002 as a simple 21-day experiment, many assorted lifelong symptoms that I attributed to stress, hard training or just 'normal' ageing vanished in that time period: arthritis in my hands, knees and hips; bowel urgency every single morning of my life; bloating and gas after meals requiring a belt buckle adjustment; frequent minor immune disturbances such as sore throats and minor colds; sensitivity to sunburn and many other annoying little issues too boring to write about (but cumulatively, a significant nuisance).

> *The most valuable Primal Blueprint recommendation is to eliminate grains, sugars and pulses from your diet. Try it for 21 days and notice the improvement in energy level, immune function, inflammatory conditions and body composition.*

Life was good before I gave up grains, but became exceptional after I made the change. Oh, I still get stressed from time to time, but it doesn't tear up my digestive system like it used to when I ate a grain-based diet. Arthritis that noticeably compromised the simple act of properly gripping a golf club (beginning around 40 years old) disappeared within months. My immune system became fine tuned such that I almost never get a sore throat or cold. There are various other complementary lifestyle practices that support my health, but I believe that eliminating grains and pulses has been the single most profound health benefit of my Primal journey. I am certain it can be profound for you as well.

As I pursue my mission of empowering people to reconnect with their genetic expectations for health, fitness and longevity, the single most valuable suggestion I can offer is to eliminate grains from your diet, and to strongly consider getting rid of pulses at the same time. If you have read this far and are still on the fence about the grain issue in particular, I urge you to conduct an experiment and discern how your body responds to a 21-day period of eliminating grains, sugars and pulses. Pay close attention to any changes in daily energy level, immune function and inflammatory conditions, and it's likely that you will experience not only fat reduction, but an improvement in many aspects of general health.

GRAINS ARE TOTALLY UNNECESSARY: SUMMARY

- Grains elicit a high insulin response, offer minimal nutritional value relative to Primal foods, and contain anti-nutrients that promote inflammation and compromise digestion and immune function. They are a cheap source of calories and entirely unnecessary for health.

- Moving away from a grain-based diet requires focus and discipline, but success can be achieved by substituting satisfying Primal foods and considering that grain foods have little taste by themselves.

- Whole grains offer more fibre and nutrients than refined grains, but potential benefits are nullified by the presence of lectins (which hamper digestion and immune function), glutens (allergenic, pro-inflammatory) and phytates (deplete nutrients). *Everyone* is intolerant at some level!

- Pulses are slightly less offensive than grains, but still contain appreciable levels of anti-nutrients and elicit an unfavourable insulin response.

- Excluding grains, sugars and pulses for 21 days can produce significant improvements in chronic health conditions, energy balance and reduction of excess body fat.

SUCCESS STORY: TIMOTHY WILLIAMS, WEST LOS ANGELES, CALIFORNIA

Timothy stumbled upon marksdailyapple.com in early 2010 and was immediately intrigued by the Primal theme due to his background as a student of anthropology and early humans. 'I was blown away by the simple premise that our bodies are finely tuned by evolution for a pre-agriculture environment,' explains Timothy. 'I read all I could about Primal living and implemented it immediately. I quickly discovered that grains had been the cause of horrible intestinal conditions I'd battled for years: irritable bowel syndrome, ulcerative colitis and maybe coeliac disease. When I ate foods such as bread, my abdomen swelled up like a beach ball. After seven years of nagging illness, I felt completely recovered after a week of eating Primally. A week!'

In 2009, prior to his discovery of the Primal Blueprint, Timothy pursued a seven-month fitness crash course. Inspired by the impending birth of his first child, the 177 cm (5ft 10in) Timothy dropped from a 96-cm (38-in) waist and over 90 kg (200 lb), to 84 kg (185 lb) and 86-cm (34-in) waist – 'through gruelling exercise and grim determination.' It was there that he plateaued, and continued to struggle with digestive issues, hormone imbalances (particularly diminished testosterone and growth hormone), recurrent tinnitus (ringing in the ears) and severe migraines requiring assorted prescription drugs.

As his gene reprogramming solidified from Primal eating, Timothy steered his fitness pursuits in a different direction, embodying a true Primal spirit in the process. Abandoning his conventional approach to working out, he says, 'I bought a pair of Vibram Five-Finger shoes and a sledgehammer, and did what came naturally to me. I had no personal trainer to goad me, no exercise classes to keep me consistent, no gym workouts under fluorescent lights.' After only two months of Primal eating and exercise, Timothy dropped from 84 kg (13 st 3 lb) to 69 kg (10 st 12 lb), boasted a 76-cm (30-in) waist, was prescription drug-free, and experienced a breakthrough in energy and vitality that Timothy likens to '... a second puberty – better than winning the lottery! Everybody in my life seems amazed at my physical transformation ... I had to throw out my entire wardrobe.'

Timothy has continued his passionate immersion into Primal living (check him out at urbanprimalist.com), bringing his family to PrimalCon, California, in 2010 and returning in 2011 to serve as a group leader and sledgehammer workout presenter. He even crafted a 'Timothy's Top 10 Most Unexpected Benefits of Going Primal':

1. **Endless energy:** On a whim, I'd find myself running shirtless for miles through winter rain after a 24-hour fast on a day I'd already exercised.

2. **Ironclad immune system:** Since going Primal I haven't had so much as a sniffle, even as my fellow office workers succumb to winter sickness left and right.

3. **No more migraines:** Who knew that my crippling migraines were a function of diet? I thought they were due to skipping meals or not getting enough sleep.

4. **Tinnitus no longer noticeable:** If you've ever had constant ringing in your ears, you know how annoying it can be, and how much you wish there were a way to make it stop ...

5. **Mad chef skills:** My whole life I resented cooking; I'd do the dishes instead. Now, with a few simple ingredients, I make things at home that are far more delicious and nutritious than the finest restaurant fare.

6. **Food bills slashed:** True, buying grass-fed meat and organic produce is more expensive. However, I eat far less frequently, derive more nutrition from what I eat and no longer need expensive, precooked, unbalanced meals and snacks.

7. **Connoisseur's palate:** I eat all sorts of stuff now that I previously considered inedible, such as kimchi, sauerkraut, sardines and chicken liver. When I eat a piece of fruit these days, it's a treat as stupendously delicious as it is rare – just as it was for Grok.

8. **Hormonal balance:** High cortisol and insulin levels suppressed my testosterone and human growth hormone levels. Today, I can feel the unmistakable effects of testosterone coursing through my veins. My muscle tone is tighter, beard thicker and I'm having to adjust to feeling particularly confident and intense all the time – an issue I welcome managing!

9. **Near-immunity to sunburn:** Throughout my life, a half-hour's direct sun exposure left my fair skin scalded, and it took weeks to recover. Now, I spend hours in the sun shirtless without even a patch of tender skin afterwards. While this is an anecdotal example, I believe the implications are profound – likely involving enhanced cell repair, cancer prevention and delayed ageing.

10. **Food is medicine:** I never suspected that the world's greatest medicines, the cures for almost every disease, the elixirs of strength and youth, are freely available in nature's most delicious foods. I used to be a ward of the medical industry, plied with expensive prescriptions and intrusive diagnostics, and told to reduce my high blood pressure by morbidly obese doctors. Now that my health is in my own hands, I am a free man at last! I don't expect to take another pharmaceutical as long as I live.

KEY CONCEPT 6:
SATURATED FAT AND CHOLESTEROL ARE NOT YOUR ENEMY

When I tell people to cut their intake of simple sugars, processed carbs and grain-based foods, they often say, 'Well, if I end up eating more fat, won't that be bad for my heart?' The answer is: no, it won't, as long as you cut out processed carbs as well as bad fats, such as omega-6 PUFAs, trans fats and partially hydrogenated oils. Eating Primally reprogrammes your genes away from a sugar-burning, inflammation-prone existence with an increased risk of heart disease, towards a healthier, leaner, stronger life with a *decreased* risk for heart disease.

Fat and cholesterol by themselves have little to do with your risk for heart disease. Conventional Wisdom's prevailing description of heart disease suggests that eating foods high in saturated fat and cholesterol, and having higher blood levels of cholesterol, cause atherosclerosis (hardening of the arteries). In fact, there's no proof at all that fat or cholesterol are the proximate causes of heart disease. We have been fed this information by well-meaning science and medical professionals who have failed to consider the role of a high-carbohydrate, high insulin-producing diet on the way we metabolise fats and cholesterol. Eating cholesterol and bad fats will contribute to heart disease *if and only if* you bathe them in a massive lifelong overdose of insulin and glucose.

> *Eating cholesterol and bad fat will contribute to heart disease* if and only if *you bathe them in a massive lifelong overdose of insulin and glucose.*

When you are a sugar burner, your excess intake of carbohydrates and polyunsaturated oils promotes oxidation and inflammation that are the real causes of heart disease. Let's reframe the story properly so you can get a basic understanding of the critical role that cholesterol and fat play in metabolic function and overall health, and how to greatly minimise your risk of heart disease, even if you have a strong genetic predisposition for it.

THE PRIMAL TAKE ON FAT

Your genes expect you to consume a variety of healthy fats as a significant part of your diet. Fats are among the most important molecules found in the human body. Cell membranes are comprised largely of fats; your brain is mostly fat (don't worry, mine is too); fat protects your organs and transports fat-soluble vitamins and we store valuable energy in the form of saturated fat.

Most dietary fats found in their true natural state – even saturated fats – are good for us, but clearly some of the dietary fats we encounter in the modern world are bad, and it's very important to distinguish between the healthy and the harmful fats. Trans and partially hydrogenated fats are among the more dangerous foods you can eat. These highly toxic fats (also known as industrial fats) are created by chemically treating vegetable and seed oils at high temperatures to render them solid. It's an inexpensive way to enhance the shelf life of all manner of processed and frozen foods. Contrary to popular belief, these agents do not even improve the flavour of food; it's all about shelf life and thus increased profits for the manufacturer, at the blatant expense of your health.

Trans and partially hydrogenated fats oxidise to form free-radical chain reactions that have been shown to damage cell membranes and other tissue in your cardiovascular system, immune system, nervous system and brain. Consumption of these agents has long been associated with cancer, heart disease, obesity, inflammation and accelerated ageing. They should simply never be consumed, which is sobering when you consider the widely cited estimate that 40 per cent of the processed/packaged/frozen/junk food items in a typical supermarket contain these ingredients.

Another class of fats that warrant concern are polyunsaturated fatty acids, also known as PUFAs. Excessive intake of PUFAs (found in industrial oils such as rapeseed, corn, safflower, and soya bean; margarine and buttery sprays and spreads and assorted baked, frozen, packaged and processed foods) can also compromise health. These fats also oxidise easily and may contribute to systemic inflammation, as the immune system tries to deal with the oxidation. They may be a major factor in arterial oxidation and inflammation. Your endocrine system is especially sensitive to PUFA consumption, which can lead to symptoms such as a slowed metabolism, low energy levels and sluggish thyroid function. PUFAs are also thought to be major players in metabolic syndrome and cancer. They should be replaced with more stable saturated fats (butter, coconut oil, lard and beef dripping) for cooking, and tasty grass-fed animal fats, sustainable fish and high-fat plants that contain monounsaturated fats (avocados, macadamia nuts, olives/olive oil). We will discuss all the best options for consuming healthy fats in Action Items 2 and 3 (see pages 121 and 129).

Trans and partially hydrogenated fats and PUFAs contain high levels of omega-6 fatty acids. Grain-fattened intensive factory-farmed meat is also higher in omega-6 fat, as are many types of nuts. While omega-6 fats offer some health benefits, we tend to consume them in excessive amounts while not eating enough of the complementary omega-3 fats. An omega-6:omega-3 (O6:O3) imbalance promotes a pro-inflammatory condition in the body. O6:O3 ratios have recently become a hot topic in progressive health circles, and it's worth making a concerted effort to get your ratio more in line with your genetic requirements for health than with the obscene imbalances produced by typical Western diet habits. Anthropologists believe that Grok enjoyed an O6:O3 ratio of 2:1 or even 1:1, while the typical Western diet ratios can commonly reach 20:1 and as high as 50:1!

Lowering your omega-6 intake and bumping up your omega-3 intake can get you more in line with your genetic expectations for robust health than the pro-inflammatory Western Diet.

While increasing your omega-3 intake with oily, cold-water fish and fish oil supplements is a sound strategy, it may be even more important to *reduce* your omega-6 intake by cutting out grains and grain-based processed foods, the PUFAs found in seed oils and chemically altered fats. We will give you other reasons why you should reduce your consumption of intensive factory-farmed animal products and most nuts except macadamia in Action Item 3 (see page 129).

QUICK FAT SUMMARY

- Healthy high-fat and cholesterol-containing foods are critical to optimal metabolic function and general health.

- The Conventional Wisdom 'lipid hypothesis of heart disease' is relevant only if you are a sugar burner, with high levels of insulin and glucose in the blood promoting oxidation and inflammation.

- Trans and partially hydrogenated fats (processed/packaged/frozen/junk foods) and PUFAs (vegetable and seed oils, margarine, buttery sprays, baked and packaged goods) disturb healthy cellular function and promote systemic inflammation, obesity and all manner of serious disease. Total elimination of these 'industrial fats' is critical.

- Emphasise intake of healthy fats, including saturated animal fats (ideally grass-fed or organic), oily, cold-water fish (high in omega-3) and monounsaturated fat plant foods (avocado, macadamia nuts, olives/extra-virgin olive oil). A healthy 'high-fat' diet (by typical Western diet standards) supports optimal hormone and cellular function, promotes satiety and raises HDL (see page 95).

- Strive to improve O6:O3 balance by eliminating bad fats, moderating intake of intensive factory-farmed animal products and nuts (except macadamia) and increasing omega-3 foods and supplements.

DAILY MEALS COMPARISON: MARK SISSON VS. KEN KORG

Following is a side-by-side comparison of a typical day of meals for me, and a typical day for a hypothetical health-conscious typical Western diet eater named Ken Korg (the modern antithesis of Grok, as featured in *The Primal Blueprint*). Of particular note is how my carb intake lands me in the Effortless Weight Maintenance zone on the Carbohydrate Curve (see page 68), despite eating a couple of hundred fewer calories than Ken. Ken's carb intake unfortunately lands him in the Danger Zone despite meal choices that are highly esteemed by Conventional Wisdom standards: an emphasis on whole grains, minimal sweets, no processed fats, and a responsible amount of total calories. Upon review of my daily meals and snacks, it's hard to argue that I've done anything but completely indulge in a variety of fabulous foods.

MARK SISSON DAILY REPORT

I was at my home for all meals on this particular day in the summer of 2011 and attempted to eat in a routine manner. However, due to the excitement of a photo shoot and the recording of data at fitday.com for this book, I think I actually ate a fair bit more than I typically do. I rarely track my food intake, but I know that I eat far fewer calories when I'm travelling or away from optimal meal choices (see pages 316–322 in *The Primal Blueprint*). On rare occasions, I'll eat significantly more calories than this, such as during our annual three-day Primal-Con retreat with catered Primal feasts in Oxnard, California.

Breakfast: Primal Omelette
4 medium pastured eggs, 2 tbsp cream, 14 g (½ oz) grated Cheddar cheese, (28 g/1 oz each) chopped mushrooms, onions, red peppers
1 cup (250 ml/8 fl oz) black coffee
(Breakfast: 30 g protein, 12 g carb, 38 g fat)

Lunch: Primal Salad
28 g (1 oz) mixed salad leaves, 28 g (1 oz) chopped onions,

28 g (1 oz) red peppers, 28 g (1 oz) cherry tomatoes, chopped chicken (85 g/3 oz), 10 g (⅓ oz) sesame seeds, 14 g (½ oz) chopped walnuts, 2 tbsp extra-virgin olive oil and lemon (homemade) dressing
(Lunch: 31 g protein, 30 g carb, 38 g fat)

Dinner: Steak and vegetables
Grass-fed rib-eye steak (212 g/7½ oz)
106 g (3¾ oz) broccoli, 85 g (3 oz) spinach, 71g (2½ oz) mushrooms, 1 tbsp butter
Glass of Cabernet Sauvignon
(Dinner: 66 g protein, 39 g carb, 38 g fat)

Snacks:
40 g (1½ oz) macadamia nuts
30 g (1 oz) dark chocolate (85% cacao):
(Snacks: 5 g protein, 20 g carb, 44 g fat)

Daily totals:
Protein: 132 grams, 528 calories, 21% of total calories
Carbs: 101 grams, 404 calories, 16%
Fat: 158 grams, 1,422 calories, 58%
Alcohol: 15 grams, 107 calories, 4%

2,461 total calories

KEN KORG DAILY REPORT

With our genes accustomed to a historical intake of around 50 grams of carbs per day or less, Ken's routine Western diet day represents six times what is genetically optimal – a glucose assault on the bloodstream.

Breakfast: Porridge and juice
60 g (2 oz) cooked whole porridge with
4 tsp of brown sugar
1 cup (250 ml/8 fl oz) orange juice
(Breakfast: 9 g protein, 50 g carb, 4 g fat)

Lunch: Sandwich, fruit and energy drink
Sliced turkey sandwich on wholemeal bread with lettuce, 1 tbsp mayo and 2 tsp mustard

600 ml (1 pint) Vitamin Water
1 large banana
(Lunch: 20 g protein, 65 g carb, 20 g fat)

Dinner: Steak, vegetables, and potato
Grass-fed rib-eye steak 212 g (7½ oz)
71 g (2½ oz) broccoli
Medium baked Maris Piper potato with 1½ tbsp butter
(Dinner: 60 g protein, 50 g carb, 33 g fat)

Afternoon snack: Energy bar (oatmeal raisin walnut flavour),
1 medium apple
Evening snack: 1 bottle light beer, 2 medium chocolate chip
cookies
(Snacks: 12 g protein, 62 g carbs, 10 g fat)

Daily totals:
Protein: 98 grams, 392 calories, 17% of total calories
Carbs: 300 grams, 1,184 calories, 52%
Fat: 71 grams, 640 calories, 28%
Alcohol: 11 grams, 77 calories, 3%
2,230 total calories

Yep, I had Ken over for dinner and fed him some delicious
steak. He opted for a baked potato (he brought it with him
in his pocket), while taking less broccoli and skipping the
other veggies. It's apparent with a few minor tweaks to Ken's
eating philosophy, he can transform to a Primal eating style
without experiencing any suffering or deprivation. The high-
sugar morning kickoff, heavy dose of sweetened beverages
(60 grams of carbs from them alone), and cravings for sugar
in the evenings, can be neutralised with some delicious, satisfying low-insulin meals and snacks.

Note: The discrepancy between the weight of foods and
macronutrient totals are due to the water content in the foods.

THE PRIMAL TAKE ON CHOLESTEROL

It upsets me to no end that the medical establishment so vilifies cholesterol. After all, cholesterol is one of the most important molecules in the body. It is involved in the structure and function of all cell membranes; the brain itself is 25 per cent cholesterol; it's a raw material for many hormones; and is the precursor to vitamin D (which is made when UVB rays from the sun react with the cholesterol near the skin surface). Cholesterol is a critical component of the bile salts necessary to emulsify and digest fats. Cholesterol is so important that we evolved an elaborate system of transporters to deliver this critical substance throughout the bloodstream to wherever it is needed.

The main lipoprotein transporters are VLDL (very low-density lipoprotein), LDL (low-density lipoprotein) and HDL (high-density lipoprotein) molecules. You are likely familiar with the oversimplified characterisations of HDL as 'good' cholesterol and LDL as 'bad' cholesterol. VLDLs are manufactured in the liver to transport mostly triglycerides (fatty acid molecules that are increased when a high-carb diet provides more glucose than can be burnt or stored) and some cholesterol to cells throughout your body. After delivering their nutrients, VLDLs shrink substantially in size and convert into either large, fluffy LDLs or small, dense LDLs. Large, fluffy LDLs are generally harmless in the bloodstream, even when levels are unusually high – something that is strongly associated with genetics.

When triglycerides are high in the blood (usually through a high carbohydrate diet and excess insulin production), VLDL production skyrockets to handle the extra load, and many of these particles can convert into small, dense LDLs. These small, dense LDLs have been identified as the problematic particles of cholesterol that can become stuck in small spaces on the walls of your arteries and later become oxidised and inflamed. This atherosclerosis process is further accelerated when you consume easily oxidised PUFAs. A diet lower in carbohydrates has been shown to reduce the number of these dangerous particles.

This is the point in the story where Conventional Wisdom goes off track. While it's true that statin drugs, or low-fat/vegetarian eating can lower the level of triglycerides and cholesterol in the bloodstream, a high insulin-producing diet can and will take whatever small, dense LDLs are still there and initiate the oxidation and inflammation process. In 2008, television journalist Tim Russert, unfortunately, illustrated the point when he succumbed to a heart attack at 58 years old despite a (statin-induced) extremely low 105 mg/dL level of total cholesterol in his bloodstream.

Conversely, HDLs are known as 'nature's dustbin trucks' because they scavenge old 'used' cholesterol in the bloodstream and return it to the liver for recycling. HDLs are very small molecules that can also easily get into the artery lining and remove small, dense LDLs lodged there, so their effect within the arterial lining is very beneficial. High blood values of HDL can be achieved with sensible (not chronic) exercise, moderating insulin production and consuming saturated fat (really!) and high antioxidant (pesticide-free) vegetables and fruits. Nailing these four recommendations goes a long way towards making you heart attack proof.

The Framingham Heart Study and Nurses in the United States Health Study, two of the largest and most comprehensive studies of diet and health ever conducted, report no correlation between dietary cholesterol intake and blood cholesterol levels, no correlation between blood cholesterol levels and heart disease and no

HEART ATTACK RISKS AND PREVENTION AT A GLANCE

RISK FACTORS

1. High carbohydrate diet: Drives excess insulin production, high triglycerides and conversion of VLDL into dangerous small, dense LDL.

2. High PUFA diet: Promotes oxidation and inflammation, allowing small, dense LDL to damage arteries.

3. Statin use: Compromises cellular energy production (depleted Coenzyme Q10/CoQ10), damages muscles and liver, and lowers HDL.

4. Exercise: Not enough promotes insulin resistance as a typical Western diet sugar burner when eating in the sugar-burning zone, or too much produces excessive cortisol and oxidative stress.

5. Genetics: Predispositions are usually only relevant when combined with adverse lifestyle (insulin, chronic exercise, stress).

PREVENTION TIPS

1. Eliminate processed carbs: Moderates insulin, lowers trigylcerides, raises HDL.

2. Eliminate PUFAs: Reduces oxidation and inflammation.

3. Increase saturated fat intake: Raises HDL.

4. Eat Primally: Moderates insulin, balances O6:O3 ratio, boosts antioxidants.

5. Exercise Primally: Raises HDL, lowers triglycerides, lowers small, dense LDL.

6. Moderate Stress: Sleep, sun, play – reconnect with genetic requirements for health!

7. Blood Tests: Focus on triglycerides; fasting blood glucose and insulin; LDL particle size (small, dense LDL), and C-reactive protein (key marker of systemic inflammation).

correlation between saturated fat intake and heart disease. These conclusions refute the premise of statin use, another example of Conventional Wisdom addressing symptoms and disrespecting causes and context (i.e. eating in the sugar-burning world).

Statins are the world's best-selling prescription drug, and arguably do more harm than good on the whole. Statins indeed slash your cholesterol total across the board (even those beneficial HDLs unfortunately) in short order, but they also produce highly objectionable side effects. Statins deplete your cells of CoQ10, a critical nutrient for mitochondrial cellular energy production. Consequently, statin users commonly experience muscle pain and weakness, liver dysfunction and chronic fatigue. Furthermore, statin use has no effect on triglyceride levels or LDL particle size, and is perhaps most effective for its marginal benefit as a mild anti-inflammatory agent. An even more profound anti-inflammatory effect, however, can be achieved through proper diet and exercise. This negates any rationale for taking statin drugs. Unfortunately, many well-intentioned people with heart attack risk factors from their familial genes and typical Western diets, and who are most in need of exercise and HDL scavenging, actually increase their mortality risk by taking statins and ignoring the root causes of heart disease.

QUICK CHOLESTEROL SUMMARY

- Cholesterol is a critical structural component of all cells and supports fat metabolism, sex hormone synthesis and vitamin D production.

- VLDLs are made in the liver to transport triglycerides and cholesterol to cells. After deliveries are made, VLDLs convert to large, fluffy LDLs (generally harmless) or small, dense LDLs (potentially dangerous).

- Small, dense LDLs can become oxidised and inflamed when insulin and triglycerides are elevated, leading to atherosclerosis.

- HDLs, 'nature's dustbin trucks', scavenge waste products in bloodstream (including small, dense LDLs) and return them to the liver for recycling. HDLs are increased by sensible (not chronic) exercise, moderating insulin production, and consuming saturated fats and high-antioxidant vegetables and fruits.

- There is no direct correlation between cholesterol or saturated fat intake and heart disease; Conventional Wisdom's lipid hypothesis of heart disease actually occurs through the lifelong overdose of glucose and insulin in the bloodstream.

KEY CONCEPT 7:
EXERCISE IS INEFFECTIVE FOR WEIGHT MANAGEMENT

Up to this point we've covered mostly diet, since these concepts require plenty of discussion and convincing in the face of flawed Conventional Wisdom. But leading a healthy, active, fit lifestyle is also of great importance, and is the topic of our final two Key Concepts before we jump into the Action Items section.

Conventional Wisdom suggests that you engage in a ridiculously unrealistic amount of chronic exercise in order to combat obesity, which is epidemic among today's sugar-burning masses. 'Sixty minutes of moderate to vigorous-intensity activity on most days of the week', is the official US Government recommendation. Of course, a sensible exercise programme balancing plenty of low-level movement with brief, intense efforts is tremendously beneficial to general health, but it's critical to understand that 80 per cent of your body composition comes from how you manipulate hormones and gene expression via the foods you eat. It's also enlightening to realise that it doesn't take much exercise to get fit, firm, lean, toned and to look good naked. If you are exercising simply to burn calories and to lose weight, you are fighting a losing battle. To put it another way, you can't exercise away a bad diet.

> *The theory of* compensation *asserts that calories burnt through exercise are* more than offset *by increased calorific intake, and generally being lazier, in the hours after exercise – a way of subconsciously rewarding ourselves for the effort.*

Recent studies assert that calories burnt through exercise are *offset, or more than offset* by increased appetite and calorific intake in the hours following exercise. This theory of 'compensation' suggests that besides the genetically driven physical craving to replenish depleted blood glucose and muscle glycogen, we also subconsciously consume more calories as a way of rewarding ourselves for working out. Furthermore, some scientists believe that structured workouts, particularly in a chronic pattern, are likely to make us lazier throughout the day. Again, the compensation principle applies on a subconscious level such that we are more inclined to take the lift instead of the stairs if we hammered out a spinning class that same morning.

I suppose you can deal with it and starve yourself after hard workouts for a few weeks or maybe even a few months, to produce impressive body composition results, as they do on *The Biggest Loser*. But unless you have your own personal trainer yelling at you constantly, or you put locks on your fridge and kitchen cupboards, exercising to manage body fat is simply unsustainable. That's why so many people following extreme regimens, under the bright lights of TV or in relative anonymity in gyms across the world, typically regain all the weight they lose during regimented sugar-burner efforts.

On the other hand, maybe you've heard that you should lift weights intensely and frequently to put on muscle, because muscle burns more calories than fat. And that's true to some extent, but not to the extent that the urban myth would have you believe. Some health professionals (Dr Oz, for example) have suggested that a pound of new muscle burns 50 extra calories in a day. If that were true, then adding 10 pounds of muscle would mean you could burn an extra 500 calories each day just sitting around the house. The problem is that a lean, muscular body doesn't burn that many more calories than a higher fat body of the same weight *at rest*. A pound of muscle burns approximately six calories per day, while a pound of fat – commonly believed to burn nothing – actually burns two calories per day.

Consider a 91-kg (14-st 4-lb) man with a basal (resting) metabolic rate of 2,000 calories a day. We know that the brain uses about 20 per cent of the calories in the body (400). The heart uses another 20 per cent (400); the liver another 15–20 per cent (300–400); and the rest of the organs another 15–20 per cent (300–400). That leaves only about 400–600 calories (20–30 per cent of the total) available for skeletal muscle. If our example man has 90 pounds of muscle, that works out to about 6 calories per pound, per day. Even if you were overweight and were to dramatically transform your body by dropping 40 pounds of fat (40 x 2 = 80 calories per day subtracted from basal metabolic rate) and gaining 20 pounds of muscle (20 x 6 = 120 calories added), you would only increase your total daily calorific expenditure by 40 calories. So it's true, 'muscle burns more than fat' … to the tune of 2 macadamia nuts per day!

> *Muscle burns more than fat … to the tune of two macadamia nuts per day!*

While exercise offers minimal benefit for weight management, there are many other wide-ranging lifestyle benefits to be enjoyed from exercise – enhanced cardiovascular, musculoskeletal, immune and cognitive function, and generally superior health and well-being. And of course, you'll look better naked. It's the movement – not the calories – that provide the benefits. Unfortunately, the sedentary forces of modern life make it very difficult to achieve the optimal level of basic everyday movement that our genes require to be healthy. And it is this basic, everyday movement that experts believe is even more critical to general health than accomplishing a vigorous daily workout and then heading off into a commute, sedentary job and an evening of digital entertainment at home. Scientists have actually coined the term, 'active couch potatoes', and have noted various health concerns, including an increased risk of heart attacks, in people who spend inordinate amounts of time sitting – commuting, at desk jobs and at home – *even if they follow a devoted schedule of daily workouts!*

For two million years, our ancestors walked, hiked, scouted, foraged, hunted, gathered, migrated, crawled, climbed and scrambled all day long. This pattern of daily activity developed a strong capillary (blood vessel) network to provide oxygen and fuel to each muscle cell, and to readily convert stored fat into energy – since fat is the main fuel used for low-level aerobic activity. Bones, joints and connective tissue became strong and resilient from all this weight-bearing, functional activity, and ageing and disease risk were neutralised.

Modern life makes it impractical, not to mention undesirable, to be out foraging for food all day. Therefore, it's critical to model the spirit of our ancestors and promote optimal gene expression by engaging in at least two to five hours per week of slow-paced movement, blending structured aerobic workouts in the proper heart rate zone (55–75 per cent of maximum, as we will discuss on page 160 in Action Item 4) with frequent, spontaneous efforts to move throughout the day. Two to five hours might pale in comparison to the hours of daily activity in the life of Grok, but will be sufficient to dramatically reduce your disease risk factors and improve all aspects of physical (not to mention psychological) health in comparison with being sedentary.

> *Basic, everyday movement and comfortably paced aerobic workouts improve fitness and fat metabolism without the burnout risk of Chronic Cardio.*

DRAWBACKS OF CHRONIC CARDIO

Unfortunately, many devoted fitness enthusiasts engage in a pattern of overly stressful aerobic workouts that are too long, too hard and conducted too frequently with insufficient rest. This approach – what I refer to as 'Chronic Cardio' – leads to fatigue, suppressed immune function, injury, failed weight loss efforts and burnout among what should be considered the healthiest and fittest modern humans.

When you hike, walk or pedal at a comfortable pace, you burn mostly fat for fuel. Your workout effectively becomes a training session for being a fat-burning beast. As exercise intensity increases beyond 75 per cent of maximum heart rate, you burn an ever-greater per centage of glucose (the preferred fuel choice when oxygen is insufficient) and stimulate the release of stress hormones into your bloodstream. Once in a while, it's fine to get out there and accomplish a 10K run or a 50-mile bike ride (if you happen to be passionate about these endeavours), and then rest appropriately afterwards. Doing it frequently without sufficient rest is where the problems start to mount.

Burning glucose and stimulating the fight or flight response promotes optimal gene expression when efforts are brief, intense and occasional, but when this happens chronically, it can lead to fatigue, depletion, sugar cravings, compromised fat metabolism and burnout. Remember, our human genes are operating on a 'survival of the fittest' principle. We are simply not adapted to grind ourselves down through chronic exercise to the point of illness, injury and burnout. For many devoted exercisers, the exhortation to 'slow down!' can lead to breakthroughs in fitness, energy levels, body composition and general health.

Exercising in your aerobic zone of 55–75 per cent of maximum heart rate will allow you to hone your fat-burning skills even further and to develop a strong fitness base without the breakdown and burnout caused by Chronic Cardio. There is a time and place to push yourself really hard and achieve fitness breakthroughs, but casual and serious exercisers alike can benefit from moderating workout pace in general. You can make your difficult 'breakthrough' workouts less frequent – and of higher quality – when you harness your resources on a day to day basis.

Comfortably paced aerobic workouts do not burn the mega-calories that chronic exercise burns, but eating right predominates over calorie concerns when it comes to weight management. Exercise is not about the calories burnt, it's about the movement, the building a solid foundation of cardiovascular and musculoskeletal fitness, and enjoying the psychological benefits of being active.

EXERCISE IS INEFFECTIVE FOR WEIGHT MANAGEMENT: SUMMARY

- Frequent medium to difficult exercise promotes the consumption of additional calories and less general activity in the ensuing hours. This 'compensation' principle asserts that exercise cancels itself out when it comes to weight management.

- Muscle burns minimally more calories than fat at rest, further negating the influence of fitness on weight management.

- Our genes require two to five hours per week of low intensity exercise for maximum health benefits and disease protection. In addition to structured aerobic workouts at 55–75 per cent of maximum heart rate, it's critical to find creative ways to move around more in daily life – avoiding the 'active couch potato' syndrome.

- Eliminate 'Chronic Cardio' patterns of excessive medium-to-difficult intensity (75 per cent of max or above) sessions that increase stress, suppress immune function, compromise weight-loss efforts and promote burnout.

KEY CONCEPT 8:
MAXIMUM FITNESS CAN BE ACHIEVED IN MINIMAL TIME WITH HIGH-INTENSITY WORKOUTS

As a complement to the need for frequent daily low-intensity movement, you must observe the 'use it or lose it' principle that fundamentally defines the process of fitness, as well as ageing. Your genes expect you to challenge your body from time to time with brief, intense workouts that help build strength, speed and power. These are great markers for anti-ageing as well as broad athletic competency. A critical factor here is to keep the intensity high, the duration short and allow for sufficient recovery between sessions.

Unfortunately, the gyms are filled with many well-meaning enthusiasts who have become socialised (that's a nicer word than addicted!) to a highly motivated, over-caffeinated, over-testoster-oned, obsessive 'no pain, no gain' goal-orientated approach. As with Chronic Cardio, a chronic approach to muscle building or any other athletic activity can lead to declining performance, fatigue and burnout. You may know people who spend hours and hours at the gym, believing that the more time they spend training, the better results they'll get. And it's not true.

On the flip side, those less inclined to fitness pursuits have become justifiably turned off with what appears to be a complex, time-consuming, exhausting approach to building muscle and getting fit. The truth is, once you understand that 80 per cent of your body composition is determined by diet, you'll also get that it doesn't require much time to become really, really fit. The Primal Blueprint

Fitness principles model the activity patterns of our ancestors to promote optimal gene expression. We need to move frequently at a slow pace, lift heavy things and sprint once in a while – it's that simple! There is no requirement to join a gym, to obsessively log miles or to subject yourself to hard-core personal trainers who can easily overstress you.

Over the course of the last few decades, I've optimised my diet and transitioned from a typical high-mileage endurance training regimen to a Primal approach. My weekly training time has declined literally 10-fold. I used to put in 20–30 hours per week of moderate to difficult sustained endurance exercise. Today, a typical week

PRIMAL BLUEPRINT FITNESS PYRAMID

The Primal Blueprint Fitness Pyramid reflects the exercise patterns that shaped human evolution for two million years. Today, honouring these simple principles promotes lifelong functional fitness, and supports a lean, toned physique – when you eat Primally.

Sprint
"All out" efforts
<10 minutes total duration
Once every 7-10 days

Lift Heavy Things
Brief, intense sessions of
full-body functional movements.
1-3x per week for 7-30 minutes

Move Frequently at a Slow Pace
Walking, hiking, cycling, easy cardio
at 55-75% of maximum heart rate for 2-5 hours per week

might include several short walks, a weekend hike of 1 to 2 hours, a couple of 20–30-minute high-intensity strength training sessions in the gym or outdoors, one sprint workout with around 7 minutes of hard effort (the whole thing lasts 15–20 minutes total), assorted mini workouts (e.g. doing some push-ups for a computer break, sprinting up a few flights of stairs just for the heck of it), and of course my beloved play time: a regular weekend Ultimate Frisbee match, practising on our backyard slackline (i.e. walking a tightrope) with my teenage son Kyle and various other spontaneous efforts lasting as little as a few minutes. All told, it's no more than 1 to 2 hours of real effort per week and a few more hours of casual movement. That's all it takes for you to become fit and delay the ageing process.

LIFT HEAVY THINGS – PRIMAL ESSENTIAL MOVEMENTS

The 21-day Total Body Transformation aims to make strength training simple, safe and appealing to exercisers of all ability levels. The last thing I want is for you to feel unqualified or intimidated by a workout recommendation that seems beyond your knowledge or ability level. Instead, I present four of the most simple – and effective – exercises ever known to mankind, the Primal Essential Movements (PEM): **push-ups, pull-ups, squats** and **planks**.

Collectively, these exercises work all the muscles in your body, and promote functional fitness for a broad application of athletic and daily life activities. These are movements our bodies have executed (in some semblance or another) on a daily, near-constant basis to promote survival for two million years. They can be done virtually anywhere with no equipment (save a bar for pull-ups), with no expert guidance or knowledge required, and with little injury risk when done properly.

> *The Primal Essential Movements – push-ups, pull-ups, squats and planks – are movements our bodies have executed on a daily, near-constant basis to promote survival for two million years.*

Since doing a sufficient number of repetitions of each baseline PEM can be difficult for many beginners, each PEM exercise offers a progression of two exercises that are less difficult than the baseline PEM, but allow you to work the same muscle groups. For example, if you can only do one or two standard pull-ups, you should scale down to chair-assisted pull-ups (double or single-legged) in order to complete an appropriate number of repetitions and build strength in the relevant muscle groups.

Once you reach the mastery level of your progression exercise, you can attempt a more difficult progression exercise. The goal is to eventually reach mastery level for each of the baseline PEMs. Once there, you can branch out into more creative and challenging exercises, all the while preserving the spirit of Primal Blueprint Fitness strength training: brief, intense, full-body, functional movements for optimal gene expression. For example, you can simply don a weight vest to safely add a considerable degree of difficulty to each PEM.

If you currently work with a trainer, do CrossFit or have a familiar and effective strength training routine, you probably don't need much guidance in this area. Go ahead and stick with what you enjoy, but please adhere to the principles of emphasising brief, intense full-body functional exercises and refrain from the all too common chronic approach to strength training. Don't hit the gym too frequently, and don't work out too long. This approach lacks the intensity and explosive action needed to stimulate optimal gene expression.

The Conventional Wisdom notion that a good strength workout lasts for an hour or more is simply unfounded by science or real-life experience. Even a 10-minute workout can produce excellent fitness benefits. Surprising as it may seem to high-tech gym enthusiasts, a couple of PEM sessions per week, lasting 10–30 minutes, can get you extremely fit, delay the ageing process and help maintain ideal body composition – when combined with Primal eating of course. If you are immersed in a hard-core routine of three, four or five 'high-intensity' workouts each week, you can actually become stronger, fitter and healthier by skipping some workouts, lessening the difficulty of some others and hitting it extra hard at those special fitness breakthrough sessions.

RUN FOR YOUR LIFE ONCE IN A WHILE – THE ULTIMATE PRIMAL WORKOUT!

The final component to Primal Blueprint Fitness is to conduct occasional all-out sprints. These brief, intense, totally Primal sessions trigger the flow of adaptive hormones that help build muscle, burn fat and increase energy levels. Sprint sessions should be conducted every 7–10 days, with the hard efforts totalling only a few minutes (e.g. a set of 6 sprints lasting 15–20 seconds each is less than 2 minutes of hard effort, wrapped into a 15–20 minute workout when you count warm-up exercises, rest intervals and cooling down). These occasional sprint workouts can have a more profound effect on your overall fitness and health than hours of Chronic Cardio. It's all about promoting optimal gene expression and challenging your body to adapt and grow stronger and faster from exercise stimulus.

While actual running is the most natural and time-efficient exercise, novices or those with high injury risk factors can choose low or no-impact options (stationary bike, cardio machines, uphill sprints, swimming, etc.) to enjoy the benefits of sprinting without the impact trauma of sprint running. Sprint workouts should only be conducted when you are 100 per cent rested and energised to deliver a peak performance. All of your workouts should align with your daily levels of energy, motivation and immune function.

Cultivate an intuitive approach to exercise with a careful balance of stress and rest. Remember, our ancestors did just enough exercise to survive, and harnessed their energy very carefully.

Instead of the robotic, consistency-obsessed approach favoured by Conventional Wisdom, feel free to apply an intuitive approach to Primal exercise. Remember, our ancestors faced harsh daily environmental circumstances and uncertain food supply. They survived by doing just enough foraging, hunting, heavy lifting and general exercise to get by, and harnessed their energy very carefully. They delivered maximum efforts when called for (kill or be killed), adapting and growing stronger throughout most of their lives (seriously, the ageing process is a modern myth), while remaining focused on a life of ease and contentment.

AGEING IS A MODERN MYTH

Much of what we think contributes to the ageing process is connected to eating in the sugar-burning zone, the excessive stress levels of hectic modern life, poor exercise habits (either chronic or insufficient) and poor lifestyle habits (lack of sufficient sleep, sun and play).

At 25 years old, I directed my genes towards the narrowly focused goal of running a fast marathon. Despite a consistent weight-lifting regimen (seriously, I was pumping iron three times a week!) and a massive daily calorific intake (between 5 and 6 thousand calories, including some 800 grams of carbohydrates), I weighed in – dripping wet – at 64 kg (10 st 2 lb) and 7 per cent body fat. I attained a top-5 performance in the US national marathon championships and a personal best time of 2:18, but the consequences of pursuing this overly stressful training regimen paralleled the ageing process: suppression of key vitality hormones such as testosterone and human growth hormone, compromised immune function to the tune of 6 to 8 upper respiratory illnesses each year, digestive problems (irritable bowel syndrome, bloating, constipation), osteoarthritis in my hips, chronic tendonitis in my ankles and knees, recurring fatigue, and ultimately physical and psychological burnout.

Today, at 58 years old, these symptoms of compromised health and ageing have completely disappeared. At 76 kg (11 st 14 lb) and 9 per cent body fat, I have added 21 additional pounds of muscle mass and a slight and healthy increase in body fat from when I was 25. My broad athletic ability has actually improved: I can lift heavier weights, jump higher, do more pull-ups and play more competently than when I was pounding the miles out day after day. No, I can't come anywhere near my marathon time from back then, but I have no desire to! On a dare, however, I could still jump into an endurance run such as a 10K and post a respectable time, thanks to the diverse benefits of my Primal Blueprint Fitness routine.

There are other examples (spend a little time Googling the remarkable photos) of athletes who seem to have defied the ageing process. Former NFL great Herschel Walker, who competes in Mixed Martial Arts at 50 years old, is visibly more ripped than when he won the Heisman Trophy at 21. Dara Torres has set longevity records in swimming, winning nine Olympic medals over a span of 24 years. As the oldest-ever US Olympic swimmer in 2008 at 41 years old, she collected three silver medals in Beijing. Her age-defying performances (and stunning six-pack) have helped her transcend swimming to become a television personality and cultural icon. The late Jack LaLanne achieved strength and conditioning performance standards as an octogenarian that represent the top 1 per cent among college-aged students, giving him claim to the 'body of a 21-year-old'.

Stuck in the paradigms of sugar-burning and flawed Conventional Wisdom, we may rationalise these examples of age-defying specimens as genetic freaks. Knowing what you know now about gene expression, you can reframe your perspective to see Herschel, Dara, Jack and, dare I say, myself, as normal and expected products of lifestyle behaviours that promote optimal gene expression. Perhaps you can see your personal potential with the expanded perspective of being able to reprogramme your genes?

I can assure you that I'm an ordinary guy who started making better eating and exercise choices over the past 25 years, and have enjoyed a consequent arresting of the ageing process. Sure, 'results may vary' due to the limitations of your familial genes, and your Primal efforts may not take you inside the mixed martial arts (MMA) cage to fight Herschel, or into the pool to battle Dara for a spot on the Olympic podium in your forties. However, promoting optimal gene expression will enable you to look and be at your best, regardless of your chronological age.

MAXIMUM FITNESS CAN BE ACHIEVED IN MINIMAL TIME WITH HIGH-INTENSITY WORKOUTS: SUMMARY

Brief, intense strength and sprint workouts promote optimal gene expression, and help delay the ageing process. Avoid unnecessary complexity or a chronic approach that compromises health.

The four PEMs – push-ups, pull-ups, squats and planks – are simple, safe, functional full-body exercises that are scalable to all fitness levels.

All-out sprints are the ultimate Primal workout! Conduct when energy and motivation are 100 per cent – once every 7–10 days is plenty. Low-impact options make sprinting accessible to all.

The ageing process is accelerated by declining physical fitness – along with poor eating, lifestyle, and stress management habits. 'Use it or lose it' and reframe your perspective about ageing by reprogramming your genes through high-intensity training.

ACTION ITEMS
THINGS YOU NEED TO DO

The following five Action Items require minimal logistics, expense and hassle. In fact, it's likely that implementing these Action Items will reduce complexity and increase flexibility in your life. While I take great pains to reject a robotic, obsessive approach in favour of an intuitive one, it's essential to make these commitments in order to enjoy the full benefits of living Primally. After reading and understanding this section, you can start your 21-day Challenge described in the next section.

Battling to control weight as a sugar burner can be such a struggle that emotional and psychological health can become compromised in the process. You will need to strengthen your resolve to leave self-destructive beliefs and habit patterns behind and commit to Primal efforts. Hopefully the immediate benefits you experience will make it easy and exciting for you to stay on track.

ACTION ITEM 1:
ELIMINATE WESTERN DIET FOODS

This is a big one, my friend. It's time to purge the kitchen and the fridge of all those foods that have caused you problems (in many cases unknowingly) in the past, and which you might default to in the future out of habit. Rather than having these items staring at you, luring you in and throwing you off track in a moment of weakness or boredom, it's best to just throw them in the bin, donate them to a food bank or even put them into storage for six months (just in case this Primal thing doesn't work out …)

Beverages:

- Designer coffees (mochas, frappucchinos)
- Energy drinks (Red Bull, Lucozade Energy)
- Bottled, freshly squeezed and refrigerated juices (acai, apple, grape, orange, pomegranate, Naked Juice, Ocean Spray, V8)
- Almond, rice, soya and other flavoured so-called 'milks'
- Powdered drink mixes (chai-flavoured, coffee-flavoured or hot chocolate)
- Powdered juice mixes (lemonade, punch, iced tea, sports performance drinks)
- Soft drinks and diet soft drinks
- Sports performance drinks (Gatorade, Powerade, Vitamin Water, Lucozade Sport®)
- Sweetened cocktails (daiquiri, eggnog, margarita)

Sweetened beverages provide heavy doses of carbohydrate without filling you up, making them among the most objectionable elements of the modern diet. Stick to water, herbal tea or coffee for your beverage choices. Sports performance drinks may be used during occasional strenuous workouts lasting over 30 minutes.

- **Baking ingredients:** Cornflour, polenta and golden syrup; starch thickeners; flours; powders (gluten, maltodextrin, milk); sweeteners (dextrose, lactose, fructose, malitol, xylitol); yeast.
- **Condiments/cooking items**: Honey mustard; jams and jellies; ketchup; mayonnaise and low-fat mayonnaise spreads; low-fat salad dressing; and anything containing HFCS; other products made with sugary sweeteners and/or PUFA oils.

Note: Using cooking/flavouring sauces such as Tabasco, soy, Worcestershire, barbecue sauce and the like may be acceptable in moderation. While these products typically contain some processed sugar and perhaps PUFA, their calorie contribution is minimal. Nevertheless, try to find the highest quality sauces with the least offensive ingredients at health food shops or on the Internet.

- **Dairy products**: Processed cheese and cheese spreads; ice cream; skimmed and semi-skimmed milk; frozen yogurt; sweetened low-fat or fat-free yogurt.

  Some dairy is acceptable in moderation if you are lactose tolerant. These include products that are raw/unpasteurised, fermented (cheese, yogurt, kefir), organic/genetically modified (GMO)/hormone-free and of the highest possible fat content (whole milk, cream cheese, cottage cheese).

- **Fats and oils:** All products containing trans or partially hydrogenated oil; buttery spreads and sprays; rapeseed, cottonseed, corn, soyabean, safflower, sunflower and all other high polyunsaturated oils; margarine; and vegetable shortening. We will discuss the fats and oils best to worst spectrum in Action Item 3 (see page 147). For now, start cooking with coconut oil, butter or other saturated animal fats and consuming extra-virgin olive oil at meals.

- **Fast food:** Burgers, chicken sandwiches, deep-fried fish fillets, chips, hot dogs, onion rings, chimichangas, tacos, chorizos, churros, and all other permutations of industrialised food we are surrounded by daily. If it has a plastic or cardboard wrapper and a Nutrition Facts panel, it's probably not appropriate for the next 21 days, nor in the future.

- **Fish:** Most farmed fish (except a few approved options, which we will discuss on page 137); fish caught by environmentally objectionable methods, or from polluted waters; large fish at the top of the food chain (shark, sword, etc.).

- **Grains:** Cereal, corn, pasta, rice and wheat; bread and flour products (baguettes, crackers, croissants, Danish pastries, doughnuts, digestive biscuits, muffins, pizza, pretzels, rolls, saltine crackers, swirls, tortillas, Wheat Thins, etc.); breakfast foods (Ready Brek®, dried cereal, eggy bread, granola, porridge, pancakes, waffles); crisps (corn, potato, tortilla, etc.); cooking grains (amaranth, barley, bulgar, couscous, millet, rye, etc.); puffed snacks, (Wotsits, popcorn, rice cakes, etc.).

    Removing grains from your diet is the number one health-boosting lifestyle adjustment you can make! Remember that corn is also a grain and not a vegetable. Corn and its derivative products (such as the particularly offensive HFCS) are ubiquitous in the modern diet and are used to sweeten all types of drinks and processed foods.

- **Pulses:** Alfalfa, beans, peanuts, peanut butter, peas, lentils, soya beans and tofu. While these foods are less objectionable than grains, they still contain appreciable levels of anti-nutrients and stimulate an excessive insulin response.

- **Meat:** Pre-packaged meat products processed with chemicals and sweeteners (breakfast sausages, dinner roasts, frozen meals, sliced meats, etc.); smoked, cured, nitrate or nitrite-treated meats (mortedella, ham, hot dogs, jerky, pepperoni, salami, etc.). Also try to limit your intake of intensive factory-farmed meats, since they are laden with hormones, pesticides and antibiotics, and contain unfavourable omega-6:omega-3 ratios due to the animals' excessive grain intake (yep, it's bad for them too). Meat is a dietary centrepiece when you eat Primally, but you must make a strong effort to favour grass-fed or organic options.

- **Processed foods:** Energy bars; fruit bars and rolls; granola bars; protein bars; frozen breakfast, dinner and dessert products; and packaged, grain/sugar-laden snack products. If it's in a box, packet or wrapper, think twice!

- **Sweets:** Brownies; sweets; chocolate bars; cake; chocolate syrup; cookies; doughnuts; ice cream; milk chocolate; milk chocolate chips; pie; sugar/sweeteners (agave, artificial sweeteners, brown sugar, cane sugar, evaporated cane juice, HFCS, honey, molasses, icing sugar, demerara sugar, granulated sugar, etc.); sugar/chocolate-coated nuts and fruit mixes; ice lollies and other frozen desserts; syrups; and other packaged/processed sweets and treats.

Consuming sweets generates a high insulin response, minimal to zero nutritional benefit, and causes an immediate suppression of immune function (since insulin competes with vitamin C at cell receptor sites). If you absolutely have to have something sweet in your house, you can acquire some high-cocoa content dark chocolate. Giving up sweets for 21 days may seem daunting, but once you clear your system of excess glucose, you will notice cravings minimise, crashes moderate and your health will improve noticeably.

ACTION ITEM 2:
SHOP, COOK AND DINE PRIMALLY

Now that your cupboards are bare and you're feeling a little shell-shocked, you can restock your kitchen with Primal foods, and implement winning strategies for shopping, meal preparation and healthy snacks. Some critics contend that eating Primally is expensive due to the emphasis on meat and organic foods, but a strategic approach can help moderate the budget impact of eating higher quality, more nutritious foods, and make meal preparation more convenient for you. In fact, several marksdailyapple.com readers relate that they actually *save* money when cutting out the expensive daily frappuccinos, energy bars and the need to consume snack foods incessantly in the carb paradigm.

> *Many Primal enthusiasts actually save money when eating Primally. They no longer require incessant snacking on expensive snacks and beverages, and enjoy metabolic efficiencies (fewer daily calories) from fat paradigm eating.*

Your Primal food options are meat, fish, poultry and eggs, vegetables, macadamia nuts, moderate amounts of coffee, high-fat dairy products, locally grown in-season fruits, other nuts and seeds, certain fats and oils, supplemental carbs for mega-calorie burners (starchy root vegetables, quinoa, wild rice) and sensible indulgences of red wine and dark chocolate. We will discuss how to choose the best foods in each category during the next Action Item (see page 129), but here we will cover shopping and meal preparation strategies to make going Primal as convenient and enjoyable as possible.

SHOPPING STRATEGIES

The majority of items in your friendly neighbourhood super-market are made with refined carbohydrates, chemically altered fats, preservatives, sweeteners and other synthetic ingredients that are not good for your health. Even the popular tip to shop on the perimeter aisles – where fresh produce, dairy and animal products are commonly found – can leave you with an unimpressive basket by Primal standards. Meat, egg and dairy choices in mainstream supermarkets are sourced almost entirely from intensive factory farming methods, but nowadays there are lots of meat, dairy and egg products that come from organic, grass-fed and free-range farms. It's time to branch out and find a superior source for your Primal foods.

Fortunately, independent health shops, farmers' markets and organic farm shops are proliferating across the country, and you may be also lucky enough to have an independent greengrocer and butcher in your nearest town. Scour your local area for superior shopping options using Google and the chamber of commerce, connecting with health-conscious eaters and visiting www.eat theseasons.co.uk, www.slowfood.org.uk and www.localfoods.org.uk to find local sources of animals and produce. Many chain supermarkets are also now jumping into the action with expanded selections of organic and locally grown produce.

There are a number of online organic retailers, which deliver organic groceries, mainly store-cupboard produce, across the UK as well as local and nationwide seasonal box schemes delivering weekly fresh organic fruit, vegetables and meat. Check out Abel & Cole, www.abelandcole.co.uk, Riverford Organics, www.riverford.co.uk and The Natural Grocery Store, www.naturalgrocery.co.uk to see if they deliver in your area.

Ideally, you would do most of your shopping at a farmers' market. Many communities host temporary open-air bazaars each week. Some communities enjoy semi-permanent venues that rotate days and locations. Perhaps you can discover a local farmer from whom you can buy produce or farm-fresh eggs directly?

Explore speciality grocery stores and markets in your area (Asian, Latino, Mediterranean, Middle Eastern, Caribbean and African) for innovative meat options, including offerings from small local farms. You may also find exotic spices and other Primal-approved fare that are uncommon in mainstream markets. Independent organic grocery stores and health food shops are typically filled with know-ledgeable, passionate, progressive-living staff members, and they make a sincere effort to source products from small, local farms.

MEAL PREPARATION STRATEGIES

You have worked hard to bring home good food, so keep the momentum going by creating an efficient kitchen. The basics would look like this: fridge freezer, cast-iron cookware, wooden spoons and spatulas (avoiding composite cookware due to chemical residue concerns) and a robust spice rack. A slow cooker or casserole dish can be useful to prepare large quantities of meat, soup, stews and chillis that can be used for delicious future meals.

Many Primal enthusiasts appreciate having a chest freezer to store larger quantities of meat and other foods. A blender is a great option for making quick Primal-approved smoothies or nut butters.

Glass tupperware and Ziploc® bags of various sizes help you store and transport food purchased in bulk, or meals made with leftovers in mind (slow-cooked soups, stews, roasts, etc.)

Primalising your kitchen and fridge will make things look much more streamlined – no more pasta, rice, flour, cereal, boxes of grain-based snacks, bottles of polyunsaturated oils, sugary sauces and condiments. Instead, your food staples might look like this:

- **Butter**: great for eating or cooking. Other animal fats such as ghee (clarified butter, great for the particularly lactose-sensitive) lard, beef dripping and even recycled bacon grease, are excellent to use for frying.

- **Coconut products:** coconut oil, milk, butter, flour, flakes and Manna™ (creamed whole coconut flesh) offer outstanding health benefits and versatile use for cooking, smoothies and to replace PUFA and wheat flour in most recipes. Coconut is high in medium-chain fatty acids, which are rare in our modern diet and offer a variety of health and disease protection benefits. Review numerous coconut posts at marksdailyapple.com for details.

- **Cookbooks**: get some creative ideas and learn how to replace virtually every typical Western diet staple with a Primal-friendly alternative. Start with the *Primal Blueprint Cookbook* and *Primal Blueprint Quick and Easy Meals*. Check marksdailyapple.com for reviews and suggestions of many other Primal/paleo-inspired cookbooks.

- **Extra-virgin olive oil:** great for salads, on vegetables and other direct consumption. Can tolerate low-temperature heating only, so do more involved cooking with coconut oil or animal fats.

- **Five favourite meals:** rotating through your favourite meals can help reduce struggles you might experience in switching to Primal eating. In Primal Essential Meals (see page 254), we provide a list of easy to prepare meals for breakfast, lunch and dinner. Primal eating can be a gourmet adventure if you are inclined, but following a basic strategy can keep things simple and within reach of even busy people who are disinclined to spend much time in the kitchen.

- **Fresh foods:** Try to eat most of the food you acquire, such as fresh produce and animal products, in less than a week.
- **Snacks:** surround yourself with Primal snacks so that you never suffer, struggle or go hungry in your switchover to Primal eating. There is a list of my favourites on page 250.

DINING OUT

For reasons both pleasurable and practical, you will likely be eating out for a small or significant portion of your meals. Becoming a modern forager requires some education and heightened awareness, but can be fun and rewarding. Remember, restaurants aim to please you so don't be afraid to throw some Meg Ryan moves into your repertoire to renegotiate the menu placed in front of you. If you're hesitant to be assertive here, feel free to embellish a little and explain that you are allergic to something – vegetable oil, wheat, gluten, processed sugar or bad attitudes from the staff. Here are a few quick tips for dining out:

- **Choose wisely:** Certain restaurants won't work no matter how hard you try to manipulate the menu. Stay away from fast food, fried food (typically made with rancid, recycled vegetable oil), inexpensive cafés and restaurants focused overwhelmingly on grains (pizzerias, etc.). Branch out into ethnic restaurants such as Greek, Indian, Italian, Japanese, Korean, Latin, Thai and even soul food. While you certainly can go wrong at just about any restaurant by choosing grain staples or meals steeped in seed oils, the aforementioned cuisines (and many others) often serve meat and vegetables in creative ways that can be a refreshing departure from your home-cooked recipes.

- **Cooking methods**: Request that your food be cooked in real butter instead of vegetable oil. This will save your body from some free radical damage and should be no hassle at all for any restaurant.

- **Macronutrients**: Build your meal around a protein source first, then consider healthy fats, then vegetables. Most restaurants have some form of animal protein that you can start with to build a decent meal.

- **Renegotiate**: Scan the entire menu and request to shuffle the deck to create your ideal Primal-approved meal. Ask to substitute extra veggies instead of the pasta that comes with the fish. Ask for extra bacon instead of the side of pancakes or toast with that omelette. Be willing to pay an extra price to trade baked potato for broccoli or have your sandwich prepared with a lettuce wrap instead of bread.

FINDING THE BEST FAST FOOD

Thanks to increased consumer awareness and demand for quality, options continue to improve for getting nutritious fare on the go. Keep in mind that animal products will be exclusively from intensive factory farming methods unless otherwise noted by progressive establishments, and of much inferior quality than what you will have at home using the aforementioned shopping and meal preparation tips. Nevertheless, when you are out in the typical Western diet world and hungry for something to eat, here are some good options and strategies for successful foraging in various food categories:

- **Chicken:** Nandos, KFC, etc. Stay away from the deep-fried stuff in favour of grilled or oven-roasted chicken. Even KFC has a griddled option on the menu (and their initials seem to have morphed from the old-time Kentucky Fried Chicken to 'Kitchen Fresh Chicken' – although the Colonel lives on). Minimise your intake of sweet sauces, even scraping the meat clean if you have to. Grab a few pieces of grilled chicken, pick a vegetable side dish if possible and be on your way.
- **Coffee houses:** Stick to straight coffee with cream and a pinch of sugar if you wish. Back off on the exotic beverages laden with sugar calories (macchiato, latte, mocha, frappuccino), and definitely ignore the baked goods (scones, muffins, croissants, biscotti) that seem to go hand in hand with coffee in the typical Western diet.
- **Ethnic:** Common ethnic fast food options include Chinese/ wok-style buffets, Mongolian barbecues and express Thai or Japanese. These cuisines make it easy to stick with meat and vegetables.
- **Hamburger chains:** McDonalds, Burger King, Wimpy. While the major global chains serve the ultimate in chemical-laden, heavily processed intensive factory-farmed meat, deep-fried fare and grain and sugar-based menu options, if you simply must partake, some clever strategies can help you stay somewhat Primal aligned. The main idea is to forgo the bun in favour of the meat. Most burger restaurants offer salad or other attempts at healthy fare. When none of this appeals, if your tour bus happens

to pull into a burger joint, it might be a great chance to stretch your legs outside and engage in an Intermittent Fast (IF).

- **Pizza:** You will have to work a bit harder to manipulate the menu at a pizzeria. Look for antipasto salads, or simply ask them to throw some chicken and vegetables (typically offered as toppings) on a plate, sprinkle some cheese, warm in the oven and serve.

- **Sandwich shops:** Subway, Pret a Manger and the like distinguish themselves with fresh ingredients and preparation. One marksdailyapple.com reader suggested ordering a plain salad, then adding vegetables and a side order or two of sandwich meat. You can also order a sandwich and request it be wrapped in lettuce instead of bread.

- **Sit-down restaurant chains:** Harvester, Aberdeen Angus Steakhouse, TGI Friday's and other national chains serving continental cuisine (burgers, steaks, ribs, salads, sandwiches, pasta and assorted deep-fried, heavily battered appetisers). Stick with meat and vegetables, and request that your meat be cooked in butter. If it's a salad you want, most of these places will add meat, chicken or fish on top for a small added charge. Avoid dressings and sauces that are laden with sugar and PUFA. Instead, request extra virgin olive oil and some vinegar.

- **Smoothie shops and juice bars:** Not much you can do here. Fruit smoothies and freshly squeezed fruit and vegetable juices deliver a sizeable sugar hit. In addition to the liquid offerings, you'll typically find an assortment of grain-based baked goods, energy bars, and snacks – another great opportunity for some Intermittent Fast (IF) if your friends are heading for a smoothie shop.

ACTION ITEM 3:
MAKE THE HEALTHIEST CHOICES ACROSS THE SPECTRUM

Transforming your diet entails that you understand the best and worst choices you can make in the various Primal food categories. Eating at the best end of the spectrum all the time can present a practical and budgetary challenge, so it's valuable to maintain the perspective that if you are eating Primal foods and avoiding grains and sugars, you are way ahead in the battle. Relax and enjoy your meals and try to get into the rhythm of staying at the highest end of the spectrum possible.

THE NEW AND IMPROVED PRIMAL BLUEPRINT FOOD PYRAMID

I'm not sure an intuitive eating style shaped largely by personal preference fits neatly into the familiar concept of a food pyramid, but it's important to counter the harmful Conventional Wisdom food pyramids with a more sensible representation of the foods that our genes require to be healthy.

Astute Primal enthusiasts will notice that my 2011 version of the pyramid has some revisions and additional detail from the original that we published with the *Primal Blueprint* book in 2009. Don't worry, we are not messing with two million years of scientific evidence, just clarifying some nuances of adapting the Primal eating style into the realities of modern life.

For example, the year-round availability of relatively high-sugar, low-antioxidant fruit warrants some moderation and selectivity in this category, particularly among those attempting to reduce excess body fat. When the wild blackberries ripen on local canes during the summer months, go for it. But reconsider that daily bowl of mangoes and pineapples year-round, particularly if you are carrying excess body fat. Grok knew nothing of these extravagances, and the insulin and fat storage impact of overeating fruit can be significant.

Herbs, Spices, Extracts
High-antioxidant/ nutritional value
Sensible Indulgences
Dark chocolate, red wine
Supplements
Multi, omega-3, probiotics, protein/meal powder, Vitamin D

Moderation Foods
Fruits Locally grown, in season, high-antioxidant (berries, stoned fruit)
High-fat Dairy Raw, fermented, unpasteurised
Starchy Roots, Quinoa, Wild Rice Athlete's carbohydrate option
Other Nuts, Seeds & Nut Butters Great snack option

Healthy Fats
Animal fats, butter & coconut oil (cooking)
Avocados, coconut products, olives & olive oil, macadamias (eating)

Vegetables
Locally grown and/or organic. Abundant servings for flavour, nutrition and antioxidants.

Meat • Fish • Poultry• Eggs
Bulk of dietary calories: saturated fat (energy, satiety, cell & hormone function) & protein (building blocks, lean mass). Emphasise local, pasture-raised or certified organic.

Pyramid Folly

Today, you can find dozens of food pyramids online sharing the common theme of eating in the carbohydrate paradigm. The original USDA Food Guide Pyramid was introduced in 1992 – an attempt to refine the 'Four Food Groups' (meat, dairy, grains, and fruits and vegetables) dogma that had shaped dietary Conventional Wisdom starting after World War II. Bowing to pressure created by escalating obesity statistics and outdated science, an updated pyramid was released in 2005. $2.5 million was paid to a private firm to create the so-called 'My Pyramid', which integrated the exercise theme and turned the food categories into 'pie' slivers, albeit with grains still dominating. Despite the inclusion of a figurine dutifully climbing stairs along the edge of the pyramid (representing the importance of regular exercise), the essence of the recommendations from My Pyramid facilitates insidious weight gain by allocating over half of daily calories to carbohydrates.

With multinational food conglomerates and the government firmly entrenched in a sugar burning mentality, advertising dollars, legislation, and funding for scientific research tends to fall in line. Today, massive profit incentives and even government subsidies drive the production and marketing of health-compromising foods such as grains, corn, soya beans, HFCS, Concentrated Animal Feeding Operations (CAFO) products, and heavily processed, packaged, sweetened, chemically altered foods and beverages. Fortune 500 companies spend $36 billion annually to market these high-margin, low-nutrition products to the public. The result is an annual obesity cost in America of $118 billion (CDC estimate in 2000), and 133 million Americans with at least one chronic disease. The simple conclusion is that Conventional Wisdom dietary policies have failed us miserably. Taking responsibility for your health entails understanding and committing to Primal eating, and striving to obtain the highest quality products in each food category.

MEAT AND POULTRY

Meat, poultry, fish and eggs will provide the bulk of your dietary calories when eating Primally. These foods are excellent sources of saturated fat and complete protein that support all facets of health, energy, weight control and peak performance. The best end of the spectrum for meat and poultry would start with a lean, high omega-3 wild animal brought down by bow and arrow in the Alaskan wilderness. Next would be a pastured or 100 per cent grass-fed animal raised in your local area. Organic animals have access to the outdoors by law, but are typically fed grains. In many cases, they may be nutritionally inferior to animals raised in pastures eating mostly grass, insects and other elements of their natural diet that boost levels of omega-3 and other nutrients.

You can typically find local, grass-fed animals at farmers' markets, local butchers or online. Beyond these premium options for meat, various other descriptive terminologies (free-range, hormone/antibiotic-free, natural/vegetarian diet, etc.) are used on products, but regulation is loose and the labelling may be of limited value in choosing the cleanest available option.

Unfortunately, the vast majority of animal products offered today are produced by intensive factory farming methods and are far inferior to a pasture-raised or organic animal. Intensive factory-farmed animals typically contain hormones (to promote faster growth and increase profits), pesticides (ingested from inferior feed sources) and antibiotics (to prevent infection from living in cramped, dirty quarters).

Due to a diet based largely on grains – as opposed to the grasses (in the case of cattle) and insects (in the case of chickens) that these animals evolved to thrive upon – intensive factory-farmed animals are generally malnourished, higher in unhealthy omega-6 fatty acids, and deficient in the more favourable omega-3 fatty acids. Furthermore, many consumers have sincere objections about the inhumane treatment and processing of these animals. Besides the health and humane objections of intensive factory farming production, the meat can taste discernibly less palatable than a naturally raised animal – bland, spongy, full of unnatural fats and devoid of rich flavour. Meat from a grass-fed animal will have a rich, flavourful taste – a difference you can taste on your first bite. Make it a high priority to find locally and humanely raised animals, or certified organic products. 'Clean' meat and poultry can cost significantly more than intensive factory-farmed meat and poultry, but it tastes better, is better for you and supports a greener, more sustainable planet.

If your budget is such that you can't afford grass-fed meats or poultry, or if you are unable to source meats at the higher end of the spectrum, take comfort in knowing that even partaking of an intensive factory-farmed animal once in a while is still preferable to a life based on cereal, bread, pasta and other sugar burner staples. When you find yourself taking advantage of that intensive factory-farmed special in the supermarket, simply make a few adjustments, such as trimming the excess fat off the cut (not necessary with a pastured animal), and using butter to cook. Take care not to over-sear the meat, for carcinogens are present in blackened meat.

VEGGING OUT

While I've conveyed the health objections of eating intensive factory-farmed meat, I will stop far short of the sensationalism and scare tactics dispensed by some anti-animal advocates. Humans evolved and thrived for two million years eating a diet centred on animal products. The potential concerns of ingesting toxins related to intensive factory-farming production can be minimised by proper cooking (don't burn meat, trim excess fat), and by eating abundant servings of high-antioxidant plant foods. While I respect the humane concerns voiced by vegetarians, and agree with those critical of intensive factory-farming production, avoiding meat is problematic from a health standpoint because too many calories will come from processed carbohydrates by default.

The health-benefits of a plant-heavy diet are undeniable, but vegetarians and vegans cannot escape the reality that health, weight control and disease prevention is still all about eating foods that our genes expect. Certainly, the risk factors for heart disease and systemic inflammation will likely be minimised if you go vegetarian instead of eating lots of intensive factory-farmed meat mixed in with excessive grains and sugars. Nevertheless, struggles with lifelong insidious weight gain, energy level fluctuations, the glucose-insulin-cortisol burnout cycle and other perils of a sugar burner metabolism are inescapable. The fact remains that no culture in the history of humanity has ever survived for a sustained period of time when eating a completely vegetarian diet.

If one must eat vegetarian, I would argue for maximum possible inclusion. Integrating oily, cold-water fish, grass-fed eggs, raw, fermented, high-fat dairy, macadamia nuts and other nuts, seeds and nut butters, high-antioxidant fruits and of course abundant servings of vegetables can make for an exceptionally healthy diet without disturbing the farm animals. The further you drift towards veganism, the greater your risk for a grain-heavy diet and the consequent insulin and micro-nutrient deficiency problems we have discussed in detail.

EGGS

Eggs can be enjoyed in abundance as a centrepiece of a healthy diet. Hopefully by now you can reject the unfounded and highly objectionable Conventional Wisdom that egg intake should be moderated (or whites should be consumed instead of yolks) due to cholesterol concerns. Egg yolk is one of the most nutritious foods on the planet, brimming with antioxidant and anti-inflammatory agents, a complete amino acid profile, omega-3 fats, saturated fats, vitamins A, E, K2 and B complex, and healthy cholesterol.

Try to consume eggs from grass-fed chickens in your local area, as they are highest on the spectrum. This often necessitates buying from a farmers' market or directly from a small farm or hobbyist with a few hens. Chickens afforded their natural omnivorous diet of insects, worms and grass can produce eggs with up to 10 times more omega-3s than conventionally raised eggs. Anyone who has tried a farm fresh egg from a grass-fed chicken can attest to the incredible flavour intensity – and distinctive orange-tinted yolk (from beta-carotene) – in comparison with a conventional egg.

Next, organic eggs are preferable to intensive factory-farmed ones since they are free of objectionable hormones, pesticides and antibiotics, and probably have less-crowded and more sanitary living conditions. However, eggs labelled organic, free-range or vegetarian are slightly nutritionally inferior to grass-fed eggs due to a diet centred on grain feed and questionable access to an outdoor, active lifestyle. Furthermore, eggs obtained through mainstream channels can routinely be 30 days old when sitting on the shelf. Being able to easily crack or peel the shell of a hard-boiled egg can indicate an aged egg with diminished nutritional value.

The budget impact of buying at the top of the spectrum with grass-fed or organic eggs is minimal, so settle for nothing less than the best you can find. For your food budget's nutritional return on investment, nothing beats £2.50 for a dozen grass-fed eggs. You can also get a little adventurous and sample something besides chicken eggs. Caviar/roe, duck, goose, ostrich, pheasant and quail eggs offer distinct tastes, excellent nutritional value and are free from the objections of eggs pumped through the intensive factory-farming pipeline.

FISH

Fish offer excellent nutritional value from complete protein, B-complex vitamins, selenium, vitamin D, vitamin E, zinc, iron, magnesium, phosphorus, antioxidants and other nutrients. Regular consumption of fish has a strong anti-inflammatory effect and can reduce heart attack risks. Due to concerns (some warranted, some overblown, as we will discuss shortly) about ingesting contaminants from polluted waters and commercial farming methods, some selectivity and moderation is warranted in order to choose the most nutritious fish.

Oily, cold-water fish from remote, pollution-free waters are some of the most nutrient-rich foods on the planet. No other food even comes close to the abundant omega-3 levels in wild-caught sustainable salmon, sardines, herring, mackerel and anchovies. FishOnline (fishonline.org) provides extensive guidance for choosing healthier, ocean-friendly and sustainable seafood, including a handy downloadable pocket guide.

> *Oily, cold-water fish from remote, pollution-free waters (anchovies, herring, mackerel, salmon, sardines) are some of the most nutrient-rich foods on the planet: no other food comes close to their omega-3 levels.*

Historically, one of the bigger concerns in consuming lots of fish was the possibility that fish at the top of the food chain could contain more mercury, a heavy metal that is toxic to humans when present at certain detectable levels. This led to the concern that people should limit their consumption of fish at the top of the food chain (tuna, swordfish, sharks, etc).

Staying lower on the food chain, favouring the least objection-able catching methods (troll or pole-caught fish over commercial longline operations), and eating a plant-heavy diet high in anti-oxidants, should alleviate most of your concerns about consuming contaminating fish. If you are especially vulnerable to toxin inges-tion (small children, pregnant or nursing mothers) or otherwise still harbour concerns, pay close attention to spectrum variables: favour-able catch methods, remote, pollution-free waters and fish lower on the food chain. For example, choosing 'light' canned tuna over white or albacore tuna will reduce mercury risks, which are already low with canned tuna since it generally comes from smaller fish than the steaks offered in restaurants or fish markets.

I suggest that you avoid most farmed fish as a general rule. Their highly objectionable production methods can compromise your health in return for minimal nutritional benefit. Most farmed fish are raised in cramped, unsanitary conditions, exposed to high levels of dangerous chemicals (dioxins, dieldrin, toxaphene and other pesticides) and fed antibiotics to ward off infections. They are much lower in omega-3 and higher in omega-6 than wild-caught fish.

There are certain categories of farmed fish with minimal toxin levels and superior nutritional profiles that make them acceptable to consume. If you are going to eat farmed fish, insist on domestic sources to alleviate the risk posed by polluted waters and lax chemi-cal regulations in high-producing countries such as China. Farmed shellfish are okay because they don't eat artificial foods and have similar living circumstances as wild shellfish – i.e. they are attached to a fixed object. Make a strong effort to eat fresh over frozen shell-fish. Catfish, crayfish and tilapia have impressive nutritional profiles and minimal toxin risks. Farmed trout from the UK is nutritionally comparable with wild trout, with minimal contaminant concerns, making it another sensible choice.

When it comes to farmed salmon, some strategic selection can support liberal consumption of this popular and nutrient-dense fish. Most of the farmed salmon available is the Atlantic species, and should be avoided due to concerns about industrial and environmental contamination, and poor O6:O3 ratios. Freshwater tank-farmed coho salmon is a superior form of farmed salmon, and even regular farmed coho salmon has acceptable omega-6:omega-3 ratios to warrant liberal consumption.

For a budget-friendly, diverse and delicious fish consumption strategy, emphasise canned sources of high omega-3 fish such as herring, mackerel and sardines. For salmon, splurge on fresh wild-caught occasionally from sustainable sources and try to find coho if you eat farmed. Supplement with other sources of wild-caught fish (ideally from remote, pollution-free waters and from sustainable sources) as budget allows and enjoy domestic sources of the aforementioned approved farmed species.

For smart shopping, cultivate a relationship with a local dedicated fish market or farmers' market seller. Fish from a speciality seller is more likely to be fresh than supermarket fare. Take aggressive, close-up sniffs of the offerings. Odour should be non-existent for fresh-water fish and perhaps a faint ocean smell for salt water fish. Bad fish will have the unmistakable smell of bad fish. Inquire about the source of the fish and stay away from Chinese or Asian imports, both farmed and wild caught. Finally, be advised that the safety, quality and sustainability of fish are in constant flux. Check resources such as the Marine Stewardship Council (www.msc.

org) and Fishonline (fishonline.org), the Marine Conservation Society's guide to buying sustainable fish, for up-to-date information.

VEGETABLES

Vegetables offer excellent antioxidant, micronutrient and anti-inflammatory properties and should form the bulk of your diet relating to portion sizes (high-fat animal products will provide the bulk of your calories, however). Adjust your mentality to make veggies a centrepiece of your meals and snacks. Get comfortable with consuming larger quantities than typical Western diet traditions call for. Be confident that it takes a whole heap of veggies to even come close to maxing out your carbohydrate 'budget' for the day. Why not enjoy an entire bowl of garden fresh tomatoes, bag of cooked spinach or head of broccoli for lunch one day? Experiment with recipes in *The Primal Blueprint* cookbooks to discover creative, flavourful, vegetable-dominant dishes.

> *Adjust your mentality to make veggies a centrepiece of your meals and snacks. Get comfortable with occasionally consuming larger quantities than typical Western diet traditions call for.*

As with animal products, try to emphasise locally grown vegetables for freshness, taste and maximum nutrient value. Local produce ranks above anything grown remotely – even remotely grown organic vegetables – because the health benefits are compromised when produce is picked early and transported long distances to your market. Most local farm operations are unlikely to use objectionable pesticides and growing methods common with large commercial operators, even if they don't assume the expense and hassle of obtaining certified organic status. Produce grown without heavy chemical influence yields higher levels of natural antioxidants (to ward off garden pests), resulting in substantially higher nutritional value and fewer health concerns than conventionally grown produce.

Make an effort to avoid conventionally grown vegetables that have large surface areas or thin, edible skins that would make them more difficult to soak and wash. For example, leafy greens (lettuce, spinach, kale) and peppers are often treated with potent pesticides that are difficult to remove before eating. Conversely, vegetables with an easily washable or non-edible skin (asparagus, avocados, onions, etc.) carry minimal pesticide risk. Try to stay away from genetically modified (GMO) vegetables, which may raise health and philosophical objections. Avoid conventional products grown remotely, since they have typically been picked too early, artificially ripened with ethylene gas during or after shipment and have nowhere near the taste or nutrient density of local produce.

You may have read how various colour groups of vegetables offer specific health benefits: reds are believed to help prevent prostate cancer, greens contribute to anti-ageing and vision, yellow and orange aid immune support and digestion and so on. It's nice to appreciate all these targeted benefits, but I prefer to focus on the big picture of abundant vegetable consumption to promote general health, and they taste great. Be wary of slipping into the 'fix it' mentality of the typical Western diet, where one bandwagon after another is rolled out on the heels of a headline study about the amazing healing properties of the latest superfood or pharmaceutical. Here is a quick list of some of the vegetables with the highest antioxidant values: beetroot, broccoli, Brussels sprouts, carrots, cauliflower, aubergine, garlic, kale, onion, peas, red peppers, spinach, swiss chard and yellow squash.

FRUITS

While fruits are excellent sources of fibre, vitamins, minerals, phenols (antioxidant, anti-inflammatory agents), antioxidants and other micronutrients, some moderation is warranted for a few reasons. First, modern cultivation and chemical treatments have resulted in fruits that are large, brightly coloured, uniformly shaped and extra sweet, with much less micronutrition than the small, varied, highly fibrous, deep-coloured, less sugary and less insulin-stimulating fruits that Grok foraged for. Second, these overly sweetened beauty queens are available all year round thanks to modern growing and transportation methods. Third, fructose (the predominant carbohydrate form contained in fruit) can cause significant metabolic problems when consumed in excess, even though it generates less of an immediate insulin spike than other forms of carbohydrates. This is particularly true if fruits are consumed in conjunction with the wildly excessive carbohydrate intake and insulin production of the Western diet.

When fructose (the carbohydrate form contained in fruit) is consumed, it's converted in the liver into not only useable carbohydrate in the form of glucose, but also into triglycerides (fat). For heavy exercisers who regularly deplete muscle glycogen, fruit is a great choice to efficiently reload their liver glycogen. However, if you ingest a good dose of fructose with your glycogen stores already full, it quickly converts into fat in the liver, and is then dumped into the bloodstream. High blood triglycerides interfere with the function of the satiety hormone leptin, causing you to want to overeat rather than rely on your stored body fat for energy. Possibly one-third of the population is fructose intolerant to some degree, evidenced by digestive symptoms such as flatulence, cramps, bloating, irritable bowel syndrome and diarrhoea. Excessive fructose consumption is also linked to fatigue, insulin resistance, diabetes and high blood pressure. Yep, eating too much fruit can make you fat.... even more so than eating too much fat, since at least fat will make you feel full!

Yep, eating too much fruit can make you fat...even more so than eating too much fat, since at least fat will make you feel full!

It's best to eat fruits with the lowest total metabolic fructose impact (see opposite) and highest nutritional value, especially if your goal is to accelerate the loss of body fat. The most simple and sensible approach is to try and emulate Grok by eating fruits only during their natural local ripening seasons. They should be grown organically and without pesticides. Minimise or eliminate your consumption of fruits grown remotely, conventionally, and out of season. Obviously, exceptions apply if you live in a fruit-challenged climate or have a strong personal preference for fruits grown tropically (bananas, pineapple, etc.) Make a concerted effort (particularly with children) to rinse off fruit with a soft, edible skin, such as berries. Fruits with thick, peeled skins (bananas, oranges, etc.) offer minimal pesticide exposure and thus less need for vigilance.

TOTAL METABOLIC FRUCTOSE VALUES FOR COMMON FRUITS

The following list, compiled from data published by *Paleo Diet* author Loren Cordain, PhD, ranks common fruits according to a 'Total Metabolic Fructose (TMF)' score that factors the impact of ingesting 100-gram servings of various fruits, after they are processed in the liver and absorbed into the bloodstream. A low TMF score implies a low impact on blood glucose/insulin production. A high TMF score implies a relatively 'high-sugar' fruit (combining a fruit's levels of glucose, fructose and sucrose). If you are interested in dialling in the best fruits for your programme, this is useful to select the fruits you can enjoy without restriction, those that warrant moderation, and those that should be avoided, particularly during weight loss efforts. The Fruit Power Rankings below ranks fruits by additional categories such as growing methods, pesticide risk and glycaemic/antioxidant values (high-antioxidant, low-glycaemic fruits being the best).

FRUIT POWER RANKINGS

This is by no means a comprehensive list, but should help you navigate the grocery store (or better yet, the farmers' market) successfully. Each list is in rank order of best to worst.

Growing Methods

1. Wild – This Grok-like fruit is difficult to find, but it's the best choice due to lack of cultivation and high-antioxidant production (think: survival of the fittest). Plant your own or hit the farmers' market!

2. Local organic – Superior choice for nutritional value, taste and safety.

3. Local conventional – Superior to remote organic due to freshness and ideal picking time. Wash thoroughly with soap or vegetable solution.

4. Remote organic – Ranks below local conventional due to harmful effects of transportation and premature picking.

5. Remote conventional – Avoid due to diminished nutritional value and pesticide risk. (Hint: if it's out of season in your area, don't eat it!)

6. GMO – Don't even think about it. 'What would Grok do?' – 'nuff said.

Nutritional Value

1. Outstanding – high-antioxidant, low-glycaemic: all berries, most stoned fruits (cherries, prunes, peaches, apricots).

2. Great – lower-antioxidant, higher-glycaemic: apples, bananas, figs, grapefruit, kiwi, pears, pomegranates.

3. Exercise some moderation (or avoid, if you are trying to lose body fat) – low-antioxidant, high-glycaemic: dates, dried fruits (all), grapes, mangoes, melons, nectarines, oranges, papayas, pineapples, plums, tangerines.

Pesticide Risk

1. High risk – soft, edible skin: apples, apricots, cherries, concentrated juices, grapes, nectarines, peaches, pears, raisins, raspberries, strawberries, tomatoes.

2. Low risk – tough, inedible skin: bananas, avocados, melons, oranges, tangerines, mandarins, pineapples, kiwis, mangoes, papayas, etc.

When characterising the Primal eating style, I'm not trying to project a 'carbophobe' vibe à la Atkins, but the numbers on the Primal Blueprint Carbohydrate Curve must be respected. If you eat abundant servings of vegetables, are somewhat indiscriminate about fruit intake, sprinkle in some moderation foods, enjoy the occasional sensible indulgence, and slip up here and there with grains and sugars, you might not gain fat if you are still eating mostly Primal foods. However, you are unlikely to lose much fat. There is not a lot of wiggle room here: shedding body fat requires that you limit your carb intake to less than 100 grams per day on average, at least until you get to your ideal body composition. It's either that or exercise incessantly like a Tour de France cyclist and eat whatever you want, or severely restrict all kinds of calories and temporarily lose muscle, water and also body fat. Neither of these alternatives is too appealing or sustainable!

Let's remember that these TMF point values are essentially splitting hairs and are not even worth addressing until you have achieved total compliance with eliminating sugars, grains and pulses from your diet. Furthermore, once you have attained your ideal body composition, you can probably consume reasonable amounts of locally grown, in season, high-nutritional value fruits such as berries, cherries and apples, even if their TMF scores happen to be high. This free licence to consume fruit is granted on the assumption that you have eliminated all forms of processed fructose from your diet, such as sweetened beverages, condiments and other hidden forms of fructose.

Very low metabolic impact (TMF score 0–2.4) – enjoy liberally at all times.
- Avocado
- Casaba melon
- Lime
- Lemon
- Tomato
- Guava

Low metabolic impact (TMF score 2.9–3.7) – moderate consumption during aggressive weight loss efforts, enjoy liberally when in season.
- Grapefruit
- Fig
- Strawberries
- Nectarines
- Apricots
- Plum
- Cherries

- Papaya
- Starfruit
- Blueberries
- Pineapple

Medium metabolic impact (score 4.1–6.3) – limit during aggressive weight loss efforts, enjoy liberally when in season unless engaging in weight loss effort, then moderate.
- Peach
- Blackberries
- Cantaloupe
- Orange
- Raspberries
- Kiwi
- Pomegranate
- Watermelon
- Banana
- Cherries, sweet

High metabolic impact (score 7.3–9.3) – avoid during aggressive weight loss efforts, moderate consumption in general.
- Pears
- Grapes
- Mango
- Apples

Very high metabolic impact (score 15.1–37.5) – avoid during aggressive weight loss efforts, and strictly limit in general.
- All dried fruits, including raisins

MACADAMIA NUTS

Macadamia nuts earn a distinction above other nuts and seeds because of their superior nutritional value. They are predominantly monounsaturated fat (84 per cent – more than any other nut or seed), which is less likely to be stored as fat, and helps raise HDL and lower LDL cholesterol. They have the most favourable omega-6:omega-3 ratio of any nut. Macadamias contain all of the essential amino acids, various forms of healthy fibre, high levels of vitamins, minerals and plant nutrients, and only trace amounts of carbohydrate. The flavonoids and tocopherols (natural antioxidants) in macadamia nuts offer excellent antioxidant properties. They have a rich and satisfying taste, making them a great snack. Macadamia nut butter is hard to find, but worth the effort and additional expense. Check out healthysupplies.com for 100 per cent raw, organic macadamia and cashew nut butters, as well as other natural products, which are delicious.

FATS AND OILS

It's especially important to honour the spectrum in this category, since good fats offer tremendous health benefits and bad fats are quite destructive. For eating, the best fats are the aforementioned pastured-raised organic animal products, oily, cold-water fish, macadamia nuts, avocados, coconut products and extra-virgin olive oil. Emphasise organic-grown olive oil with the 'first cold-press only' distinction.

If you are not a big fish eater and wish to supplement your omega-3 intake, you can find refrigerated omega-3 oils in dark containers in the health food store, but look for something other than flax oil. While flax is the most common offering and is high in omega-3, recent research suggests that the predominant type of omega-3 found in flax oil, alpha-linoleic acid (ALA), is difficult to assimilate in the body. It must be enzymatically converted into the more useful omega-3 fractions: docosahexanoic acid (DHA) and eicosapentanoic aicd (EPA). Even then, there's still no guarantee that your body will handle that conversion efficiently. Some alternative offerings you might find are borage, cod liver and hemp seed. Pharmaceutical-grade omega-3 fish oil capsules, high in DHA and EPA, offer an excellent budget-friendly alternative to boost omega-3 levels.

For cooking, saturated fats are more temperature stable than PUFAs, so you can heat them to high temperatures without risk of oxidisation, and resulting damage to cell membranes. Coconut oil, the most saturated of all vegetable oils, is a great choice for cooking. Animal fats such as butter, ghee, lard, recycled bacon grease and beef dripping are all excellent options for cooking. Isn't it fascinating how so many of the things we were told were bad for us are actually very healthy?

Make a strong effort to completely eliminate trans, partially hydrogenated and PUFA fats from your diet, which should happen automatically when you reject heavily processed, packaged, frozen and fried foods (that are often made with grains as well). Replace PUFA products such as rapeseed oil, soya bean oil, safflower oil, corn oil and margarine spreads with butter and extra-virgin olive oil for eating, and coconut oil or animal fats for cooking.

MODERATION FOODS

While I respect the hard-core Primal/paleo eaters who apply stringent guidelines to reject all 'modern' foods, I believe it's more important to make the Primal eating style as enjoyable and accessible to as many people as possible. These categories of foods are acceptable in moderation in the Primal eating style.

- **High-fat dairy products:** While dairy should not be a centrepiece of your diet, if you are lactose tolerant and feel a need to keep some dairy on your programme, you can enjoy certain dairy products at the highest end of the spectrum. The best choices are raw, fermented, unpasteurised, unsweetened and high-fat options such as ghee, butter, full cream, aged cheese, cottage cheese, cream cheese, Greek-style full-fat yogurt, kefir and raw whole milk. Stick to pasture-raised/grass-fed or organic dairy products to avoid the hormones, pesticides and antiobiotics common in commercial dairy products. Eliminate fruit sweetened yogurt, frozen desserts and other high-carb dairy offerings. Stay completely away from regular pasteurised, homogenised, semi-skimmed and skimmed milk.

The best choices in dairy are raw, fermented, unpasteurised, unsweetened and high-fat products from pasture raised/grass-fed animals.

If you can source them in your area, raw and unpasteurised dairy products are recommended. That's right, some exposure to beneficial bacteria from raw foods helps strengthen your immune system, while the pasteurisation process greatly compromises a product's nutritional value. Dairy products made by intensive factory farming methods require pasteurisation due to the high risk of food-borne illness in crowded commercial plants, which is why it's best to choose alternatives made in cleaner environments. Fermented dairy products offer exposure to beneficial probiotics, promoting digestive health and decreasing cancer risk. Fermentation also breaks down lactose, reducing the carbohydrate content and preventing digestive problems in sensitive people.

Giving up conventional low-fat dairy products relieves you of lactose sugar (difficult to digest and stimulates excessive insulin production), casein (an objectionable protein with immune-compromising properties), epidural growth factor (EGF increases risk of cancer and tumour growth), recombinant bovine growth hormone (rBGH increases milk production but makes cows, and humans, sick) and other hormones, pesticides and antibiotics. Furthermore, the calcium benefits of dairy are overblown. A typical grain-based diet can grossly interfere with calcium absorption, making people think they need even more calcium.

Ultimately, typical Western diet eating can actually result in excessive calcium intake, altering the critical balance of calcium-to-magnesium in the body. It's more worthwhile to focus on increasing magnesium intake from leafy greens, nuts, seeds and fish. For bone health, getting enough vitamin D from sunlight is exponentially more beneficial than gulping down a glass of milk every day. A 225 ml (8 fl oz) glass of fortified milk offers around 100 I.U. of vitamin D, while 20–40 minutes of direct sun exposure in a swimsuit offers around 10,000 I.U.

- **Other nuts, seeds and their derivative butters:** Nuts are fairly good sources of protein, fatty acids, enzymes, antioxidants, and vitamins and minerals. Their incredible nutrient density allows you to consume fewer calories to satisfy your hunger and energy needs compared to grain-based processed snacks. But be careful: some people find that they overdo the nuts on a Primal Blueprint eating programme, especially as snacks, and take in more calories than they may have guessed. When you ingest significant fat calories from nuts, your body won't look to your stored fat as easily for energy. Also, some moderation is warranted in order to achieve a favourable overall dietary O6:O3 ratio. For example, walnuts are lauded for being a great source of omega-3s (highest of any nut or seed), but they are five times higher in omega-6 than omega-3.

Pumpkin seeds have the most favourable ratio among seeds, with significant values of omega-3s and 'only' double that level in omega-6. Enjoy various other nuts and seeds as salad toppings or snacks, but emphasise calories from macadamia nuts, vegetables and animal foods when replacing calories previously obtained from grain foods. Also keep in mind that peanuts are actually pulses, not nuts, and should be minimised or avoided since they are highly allergenic and can develop moulds that produce afla-toxin, a potent carcinogen.

Nuts are good for about 6 months (12 months if still shelled) and are best stored in a cool, dry place. Use the fridge or freezer if you don't plan to eat them quickly. If your nuts have a rancid, oily smell or any discolouration, discard them immediately. Nut and seed butters are a great snack option. Try to find raw, cold-processed butters that are simply ground up (at low temperatures and free of added ingredients – except salt, which is fine), and refrigerate them at all times.

- **Supplemental carbs:** Sweet potatoes, quinoa and wild rice offer the most healthful source of carbohydrates for heavy exercis-ers who require additional dietary carbohydrates to replenish frequently depleted muscle glycogen. The select few who fall into the 'need supplemental carbs' category are effortlessly main-taining ideal body fat levels, and train so heavily that a sudden and severe drop in energy and delayed recovery are occasionally concerns when eating Primally. Think Tour de France cyclists, NBA basketball players or cross-country runners.

For the rest of us, instead of obsessing on 'reloading' with carbs as Conventional Wisdom has talked about for decades, it is better to prioritise obtaining the adequate number of dietary protein grams for your body weight and activity level. Meeting your protein needs will help preserve or build muscle tissue, and also facilitate gluconeogenesis (to replenish muscle glycogen and keep blood glucose stable), even when dietary carbohydrate is limited.

If you decide to eat more carbs, sweet potatoes are superior to the excessively starchy white, Maris Piper, Rooster or new potatoes. Quinoa is technically not a cereal grain, but a relative of leafy green vegetables including swiss chard and spinach. Vegetarians (and even Primal enthusiasts who harken for a grain-like taste or recipe base) laud quinoa for being a complete protein (containing all 9 essential amino acids, with 12–18 per cent of total calories as protein) and free from gluten.

Wild rice is also not a cereal grain, but an aquatic grass. It offers a better nutritional value than cereal grains, a nearly complete profile of essential amino acids (14 per cent protein calories) and is gluten free. Quinoa and wild rice are the best options if you must have a grain-type experience. Categorise these supplemental carbohydrate foods as an indulgence. They may be enjoyable, but are probably unnecessary, especially if you are trying to reduce excess body fat.

Unless you follow an extreme training programme with heavy refuelling requirements, supplemental carbs should be considered indulgences; they're unnecessary for health, especially if you are trying to reduce excess body fat.

BEVERAGES

After purging your fridge of sweetened drinks you are not left with many options, but you will also be giving your system a nice break from the insulin surges prompted by the juices, smoothies, fizzy drinks and sporty concoctions consumed in excess in the typical Western diet (not to mention HFCS, bone-depleting phosphoric acid and other chemicals and preservatives contained in many processed drinks). Fruit and vegetable juices, even fresh from the juicer bursting with antioxidants, are not necessary. The large dose of sugar and resultant insulin surge overshadow any purported super-nutrition benefits. I'd prefer you focus on consuming whole food sources of produce instead of juicing.

Water will become your must-have beverage, but it's important to reject the groundless Conventional Wisdom notion that you must robotically drink eight glasses per day to be healthy. This assertion is completely unsupported by science, and can potentially lead to some digestive and health problems associated with excessive intake. In fact, your thirst mechanism and your kidneys do an excellent job at regulating your hydration levels, and every drink and solid food you eat contributes to hydration to some degree. Rather than worry or obsessively track your water intake, simply let your thirst be the guide.

Conventional Wisdom's recommendation to drink eight glasses of water per day is unsupported by science. Simply let your thirst mechanism guide you to optimum fluid intake.

Certain other beverages are Primal-approved. Soda water or mineral water can add some pizazz to an ordinary glass of water and help assuage the elimination of a fizzy drink habit where carbonation is part of your comfort ritual. Add a squeeze of lemon, a handful of berries and a pinch of sea salt for some extra taste. Teas can be a great way to boost antioxidant intake and spice up drinks. Potent 'true' teas (white, green, black) offer anti-inflammatory and immune supporting benefits, protecting against cardiovascular disease, osteoporosis, arthritis and maybe even some cancers. For all varieties, go for the freshest leaves possible and use loose leaves or sachets over bags if feasible. You may find it convenient to make large batches of tea and chill them for later use.

Coffee may not have been part of Grok's daily routine, but I enjoy my daily cup of coffee and see no reason to avoid it for health reasons. Since caffeinated coffee is a mild central nervous system stimulant, it's important to refrain from using coffee as a crutch to prop up sagging energy levels, and to moderate your overall caffeine intake to avoid becoming habituated. Instead, choose decaf for your second cup of the day, engage in good sleeping habits, moderate insulin production in your diet, exercise Primally and boost energy naturally with cold water plunges, deep breathing sequences, napping or quick exercise breaks after long periods of inactivity. Make an effort to find organic, fairtrade coffee due to questionable chemical and pesticide use in some major coffee-producing countries. Darker roasts contain less caffeine, are more flavourful and offer more health benefits.

HERBS AND SPICES

Herbs and spices are an important element of the Primal eating style due to the variety of flavour and nutritional benefits they offer. Herbs are green plants or plant parts used to add flavour to foods.

Spices are used to enhance flavour, add colour or help prevent bacterial growth on food. Herbs and spices support cardiovascular and metabolic health, may help prevent cancer and other diseases and improve mental health and cognition. Some of the highest antioxidant values of all foods can be found among herbs and spices.

When used in cooked meat dishes, herbs and spices also reduce fat oxidation and the formation of potentially harmful compounds. Both dried and fresh herbs and spices offer excellent antioxidant values. Try to find organic sources for dried herbs, since conventional products are usually irradiated (diminishing nutritional value) and pesticide residues are easily ingested. Here are some flavourful herbs and spices with excellent antioxidant values that should be used regularly in your kitchen.

- Basil
- Black pepper
- Cayenne powder
- Chilli pepper
- Coriander
- Coriander seeds
- Cinnamon
- Cloves
- Cumin seeds
- Dill
- Ginger
- Mint
- Mustard seeds
- Nutmeg
- Oregano
- Paprika
- Parsley
- Peppermint
- Rosemary
- Sage
- Tarragon
- Thyme
- Turmeric

Regarding salt, Primal eating will default you into an optimum intake range for sodium. Cutting out processed foods will remove several thousand milligrams of sodium from your diet, and consequently diminish your cravings for more. If you enjoy the flavour of salt, go ahead and shake a reasonable amount on your food without trepidation. There is actually a very minimal correlation between sodium intake and high blood pressure or heart disease. Your best choice is sea salt, which is richer in minerals and trace elements than table salt, which has been nutritionally diluted by the refinement process.

SENSIBLE INDULGENCES

Sensible indulgences honour the Primal Blueprint philosophy of enjoying life and not stressing about perfection. They can enhance your appreciation of food and the celebration of life's great moments with friends and family. To fully appreciate an indulgence, you must adopt a guilt-free mentality, focus intently on the pleasure of the experience and of course observe a moderate approach. After all, too much of a good thing is not good at all.

- **Alcohol**: I could officially recommend that you don't drink alcohol at all, since the ethanol part of alcoholic drinks is a potent toxin. However, limited intake of certain alcoholic beverages should not harm you, and may actually give you some antioxidant benefits. Obviously, drinking moderately and responsibly is critical. Your 'when to say when' point can be ascertained by common sense, gender and body weight, historic tolerance level and environmental influences (Do you have to drive home? Are you tired, stressed, hungry or otherwise potentially more susceptible at this particular time?).

  Red wine is the superior alcoholic beverage choice for its impressive antioxidant benefits. Beer is marginal (after all, it's made from grain!), while spirits and mixed drinks using sweetened beverages are the ones to be avoided. Note that alcohol calories (at 7 per gram) are devoid of nutritional value and are generally the first to burn when ingested, which means burning stored body fat is interrupted while you indulge.

- **Dark chocolate:** If you have a sweet tooth that is challenging your Primal transition, prioritise dark chocolate as a sensible indulgence. It has one of the highest antioxidant values of any food, along with brain-stimulating compounds, euphoric agents and satisfying saturated fat. The key here is to find products with the highest possible cocoa content. At least 75 per cent is excellent, but as you get closer to 100 per cent cocoa, the lack of sweetness will be make this indulgence tough to appreciate (85 per cent is perfect for me). It may take a bit of time, but once you get used to the rich taste of high cocoa content dark chocolate, you will likely lose your affinity for milk chocolate, which will suddenly seem too sweet. Try to find organic chocolate due to pesticide concerns with cocoa bean production in foreign countries.

GOING PRIMAL ON A BUDGET

One of the criticisms that really bugs me is when people claim Primal eating is prohibitively expensive. My gut reaction is 'expensive relative to what?' Consider the costs of obesity/metabolic syndrome care (£77 billion a year in the USA), cancer treatments or the prolonged use of NSAIDs (non-steroidal anti-inflammatory drugs – heavy use can run over £131/month), statins (£33–131/month) and other popular prescription drugs used to combat health problems heavily influenced by poor dietary habits.

Now that I've calmed down a bit, how about a deeper analysis of the typical Western diet food budget, with processed and fast foods, sweetened drinks, designer energy bars, meal replacement weight-loss products, frozen and packaged treats, and the full complement of grain-based snacks to jack up the register totals? Going Primal will eliminate significant budget expenses for these foods and switching over to the fat paradigm will likely reduce the number of daily calories you require to sustain energy (see Timothy Williams' Success Story on page 84).

Granted, even if you eat less as a modern forager, it's likely going to be higher cost produce. Pastured-raised or organic animal products can cost up to twice as much as conventional fare, while increased produce costs are not quite as severe but still significant. For argument's sake, let's say your overall food budget increases somewhat when you eat Primally. Can you reflect on the bigger picture, weigh the relative importance of various other discretionary expenses and perhaps increase the priority of your food budget a bit? Now that I have you in the proper frame of mind, let's explore a few ways that you can eat Primally on a budget:

- **Alternative shopping:** Farmers' markets are at least competitive with major store prices, and great bargains can be had around closing time when sellers are packing up. Anne Brown, a marksdailyapple.com reader in California, reports that her town's speciality meat store (offering vastly superior local meats) has prices competitive to the local chain supermarket – likely due to reduced transportation, overhead and advertising costs.
- **Home grown:** Plant some seeds and get your garden plot going! It doesn't get any cheaper or more fun than growing your own. Besides generating convenient, nutritious, inexpensive fresh produce, growing your own provides the psychological benefit of strengthening your connection to the food you eat in the age of industrialisation. The same is true for hunting the animals you

eat, if that's your thing. Check out marksdailyapple.com for posts on 'urban gardening'.

- **Storage**: Invest in a chest freezer and avail yourself of discounts for buying in bulk – either from teaming up with friends and neighbours, big box stores, the Internet or other sources.

- **Work**: Put in some part-time hours at a local co-op or alternative grocer to enhance your food education and enjoy employee purchase discounts typically ranging from 10–25 per cent.

HONOURING THE 80 PER CENT RULE

Navigating the spectrum can get a little complex and potentially intimidating, so this is a good time to discuss one of the founding principles of the Primal Blueprint, the **80 Per Cent Rule**. The essence of this rule is do the best you can to stay aligned with healthy living, eating at the highest end of the spectrum and promoting optimal gene expression, but not to stress out about perfection. When you establish a foundation of healthy habits, your body can become quite resilient to the occasional late night of revelry, indulgent slice of cheesecake or long night in front of the computer working on a deadline project. Living Primally is about waking up the next morning, recalibrating your compass and getting back on course to honour your genes after occasional detours.

To be clear, I'm not offering a free pass to leave grain meals and chronic cardio workouts intact in your daily schedule. The spirit of the 80 Per Cent Rule is to strive for 100 per cent compliance, and accept an 80 per cent success rate due to the pressures, distractions and logistical challenges of living Primally in the modern world.

ACTION ITEM 4:
EXERCISE PRIMALLY – MOVE, LIFT AND SPRINT!

The Primal Blueprint Fitness plan is within reach of everyone, from sedentary people just embarking on fitness pursuits, to reformed Chronic Cardio athletes and gym enthusiasts looking for a more sensible, time-efficient and fun way to get fit, or even superfit. The basic framework of Primal Blueprint Fitness is to move as much as possible at a low-level pace, both through structured aerobic workouts and general efforts to increase daily activity (e.g. the stairs instead of the lift), conduct two high-intensity strength training sessions per week lasting 30 minutes or less and conduct a brief, all-out sprint workout once every 7–10 days.

MOVING FREQUENTLY

You learnt in the Key Concepts that exercise isn't about calories burnt, but about movement. It's imperative now to discover ways to simply move around more – even if only for brief periods – throughout your day. Make it an official personal policy to take the stairs instead of the lift, park at the furthest spot in the car park instead of always angling for a closer one, and generally prioritising pedestrian movement over sedentary options. Here are several tips to add more movement to your daily routine:

- **Wake-up stroll:** Grab the dog and take a lap around the block to gradually build energy and prepare for a busy day. Even if you only have 5–10 minutes to spare in the morning, it's well worth the effort.

- **Brief work breaks:** Mounting evidence suggests that work productivity, mental health and stress management can improve significantly when you moderate digital stimulation and take frequent breaks away from focused, sedentary tasks to engage with fresh air, sunlight, open space and physical movement. Get outside and stroll around the office courtyard, up and down the building stairwell or otherwise make do with whatever your surroundings. When you sit back down at your desk, you will have a perceptible improvement in energy and focus.

- **Stroll before arriving home:** When you pull into the driveway after a long day behind the desk, behind the wheel and behind in paperwork, hit the road for 5–10 minutes before you kill the momentum by opening the front door. You may be justifiably tired and dreaming of a cold beer and a soft couch, but you will get energised and refreshed by a simple stroll.

- **After-dinner stroll:** This is a great one to involve the entire family. Clear the plates, throw on some Vibram Five Finger® shoes (to simulate a barefoot experience – check out vibramfivefingers. com), and take the dogs and/or humans on a spin around the block. Regular 10–20-minute outings will establish a wonderful tradition of winding down the evening, and present an appealing alternative to going from the dinner table straight into the TV or computer room. Take some deep breaths, share some conversation and return to your abode with a perspective shift that there is more to the evening of leisure than digital media.

- **Grand weekend outing:** Take a walking trail at a National Park, or an urban journey to the farmers' market and back home. Set a reasonable goal based on your existing fitness level. Nearly everyone can walk or cycle for at least an hour at a comfortable pace, and many can enjoy a two or three-hour hike.

AEROBIC WORKOUTS

An aerobic workout is defined as anything that elevates your heart rate into the target zone of 55–75 per cent of maximum for a sustained period of time. You can choose activities that are most appealing to you: brisk walking (or jogging if you are fit enough to stay under 75 per cent), easy cycling (on a stationary machine or cruising around town for errands), various cardio machines at the gym or winter/water activities such as snowshoeing, cross-country skiing, swimming, rowing and stand-up paddling (shameless plug for one of my favourite pastimes). Use some caution when doing a group exercise class or outdoor workout, because lively instructors, pulsating music and enthusiastic training partners can easily elevate your heart rate out of the aerobic zone and promote a chronic exercise pattern.

Exercising at 55 per cent of maximum feels really easy and most people will get here while walking around the block. Seventy-five per cent of your max heart rate still feels quite easy; you can carry on a conversation without getting winded and end your session feeling refreshed instead of depleted. To pinpoint your zone numbers, you can perform a strenuous maximum heart rate test (get medical clearance first), or use a formula to estimate your max. The long-standard '220 minus age' formula is now believed to produce material inaccuracies in some people, so we recommend a new formula from University of Colorado researchers as follows: 208 minus (0.7 times age) = Estimated Maximum Heart Rate. For example, a man aged 40 has an estimated max heart rate of 180 beats per minute (208 – 28 (0.7 x 40 = 28). His aerobic zone upper limit would be 75 per cent of 180, or 135 beats per minute.

> *Exercising in the aerobic zone of 55–75 per cent of*
> *maximum heart rate should feel quite comfortable, and*
> *leave you feeling refreshed and energised after workouts.*

Use a wireless heart rate monitor (options abound starting at £30; Polar is the leading brand), for the most direct and accurate feedback of your heart rate during exercise. Otherwise, you can calculate your training heart rate by pausing during your workout and placing your finger against the carotid artery on the side of your neck, the best place to feel a strong pulse. Check your watch and count how many beats occur in exactly 10 seconds then multiply that number by 6 to determine your heart rate in beats per minute. That's how I do it!

Please make an effort to regularly monitor your heart rate after you get into the rhythm of your workout, particularly at times when you encounter a challenge such as a hill (or an obnoxious poser passing you abruptly on the bike path) that would necessitate some restraint. Going by perceived exertion alone can cause you to exceed the recommended zone because a 75 per cent effort feels so moderate, and in many cases is well below your typical workout intensity level.

Don't worry too much about monitoring the lower end of 55 per cent of maximum heart rate. Super fit people might have to escalate their walking pace a bit to reach 55 per cent and achieve the desired aerobic training effect, but most of us will reach that minimum level simply by putting one foot in front of the other. There is also a subjective component of proper aerobic workout intensity to respect. Aerobic sessions are designed to energise and refresh you. If you feel at all taxed or detect a craving for sugar after a session, it's a safe bet that you have drifted outside your aerobic zone and compromised the desired effect of the workout.

During the 21-day Challenge described in the next section, a sensible pattern of Primal workouts is suggested, including a total of eight (if all goes well) aerobic workouts of either extended duration or moderate duration. If 'moderate duration' represents your typical session (say, between 20 and 45 minutes for most exercisers), an 'extended duration' workout is something significantly longer that stretches your fitness capabilities. These are parameters that you will specify based on your workout history. As your fitness progresses, you can lengthen the duration of both your moderate and extended duration workouts, while of course taking care to keep your heart rate in the proper range.

STRENGTH TRAINING SESSIONS

Primal Blueprint Fitness is predicated on performing 'functional' strength movements – exercises that use many muscle groups at the same time for efforts that have broad, real-world application. There are four PEMs that form the basis of this training style – **push-ups**, **pull-ups**, **squats** and **planks**. PEMs are simple movements that can be done by exercisers of all ability levels, in a gym, at home or adapted to include equipment if desired. During your 21-day Challenge, you will conduct a total of seven PEM workouts, either full length (30 minutes) or abbreviated (10–15 minutes) as directed.

A full-length session might entail two or three sets of maximum reps for each of the four aforementioned PEMs. An abbreviated session might entail a single set of maximum reps for each PEM. Rest just enough (30–60 seconds should do it) between each exercise to allow your respiration to return to near normal. If your muscles fail before reaching the goal number of repetitions, rest for 5–10 seconds and then try to add a few more reps to your total. However, don't linger too long on any one exercise. Do as many reps as you can and move along quickly.

All workouts should have a brief warm-up period of 2 to 5 minutes of aerobic exercise to elevate body temperature and transfer blood from your torso to your extremities in preparation for intense efforts. Try some gentle jumping jacks, a few stair climbs, jogging (either in place if indoors, or loop around your outdoor venue) or using a stationary bike. After exercise, a gentle cool-down session of a few minutes of light aerobic exercise and perhaps some basic Primal stretches such as the Grok Hang and the Grok Squat (see page 172) is advised.

KEEP IT SIMPLE AND PERSONAL

I'd like to emphasise that the Primal approach to strength training is one of simplicity and flexibility. Fitness experts might debate various 'right' ways to strength train, but I like to focus on the big picture view of developing lifelong functional fitness, enjoying yourself during workouts, tailoring your fitness endeavours to your lifestyle and of course promoting optimal gene expression.

At marksdailyapple.com, we publish a Workout of the Week (WOW) with creative interpretations of the PEMs, and hundreds of readers contibute with suggestions to keep workouts fun and interesting. At some workouts, I'll cycle through the PEMs in a linear fashion, while other days I may focus on a single exercise – such as push-ups – and try to get 200 or 300 in over the course of 30 minutes, or during three 10-minute sessions over the course of the morning hours. Sometimes I'll do a set of PEMs, run a few wind sprints, do another set, run a few more sprints, etc., to complete a challenging workout. It all works!

If you are a creature of habit and prefer to do the same exact workout over and over, this is perfectly fine and will deliver exceptional fitness benefits. The popular notion of 'confusing' your muscles with a never-ending stream of new exercises is not secretly super-effective, nor is it inherently harmful. Let your workout choices be guided by personal preference, making sure to honour the 'brief, intense' Primal philosophy. Of course, safety and competency with the movements you choose is also critical, which is why I recommend such a simple and scalable set of exercises.

PRIMAL ESSENTIAL MOVEMENTS

Following are brief descriptions and illustrations of each PEM and the progressions up to and beyond the baseline PEM. For more details on the PEMs, please visit marksdailyapple.com for video instruction and a free ebook, *Primal Blueprint Fitness Guidebook*, which you can download. This guidebook has detailed descriptions of additional progression options for each PEM, and numerous advanced variations. However, the material here should be all you need to build a custom-designed total body strength training programme, whether you are sedentary or quite fit.

Men – Essential Movement mastery: These mastery levels represent one set of maximum effort. If you fall short of any of these benchmarks, drop down to the appropriate progression exercise to build up your strength for a future attempt at reaching mastery numbers.

- 50 Push-ups
- 12 Pull-ups (overhand grip)
- 50 Squats (thighs parallel to ground)
- Plank: 2 minutes holding the Forearm/Feet Plank position

Women – Essential Movement mastery:

- 20 Push-ups
- 5 Pull-ups (overhand grip)
- 50 Squats (thighs parallel to ground)
- Plank: 2 minutes holding Forearm/Feet Plank position

Mastering each of these PEMs indicates an excellent level of functional fitness and broad athletic competency. In combination with plenty of low-level aerobic exercise and occasional sprints, your PEMs efforts will enable you to play hard with minimal injury risk, participate in a variety of physical challenges and put the finishing touches on a lean, toned physique (remember though, 80 per cent of your physique depends on your diet). Furthermore, you will neutralise the ageing process as we perceive it, which is strongly associated with declines in physical strength, power, endurance, muscle mass and – by consequence – organ function.

Your first PEM workout will be an assessment session to determine the appropriate starting exercise in each PEM progression. This may take more than 30 minutes with all the trial and error involved. Review the progression exercises and start with the exercise that you think you can complete the appropriate number of reps of. If you over- or undershoot, choose the most appropriate exercise and work there for 2 to 3 sessions. When you can increase the number of maximum reps by approximately 25 per cent (e.g. going from 9 chin-ups to 12 chin-ups), bump up to the next PEM progression exercise. Remember, the goal of PEM sessions is to challenge your muscles to their failure point. I believe that even a novice can achieve Essential Movement mastery in a few months with diligent effort.

Back slide arches: Sit on ground with legs extended and back perpendicular to ground. Press down with arms and arch back until entire body is straight. Before landing, drive bottom backwards through the air and land further down the field in a sitting position, still facing backwards. Once reaching the 25 or so meter mark, face forward and extend body into hand/feet plank position.

Push-ups: Assume plank position with arms extended at shoulder width, and hands facing forwards on the ground. Lower carefully until chest touches the ground, while preserving a straight body position from head to toe. Elbows will bend towards the ground at a 45-degree angle backwards from the starting point. Keep core and glutes tight, and head and neck neutral (aligned with torso) throughout sequence. Make sure chest touches first!

Push-up progressions: Baseline Essential Movement mastery: men 50, women 20
- Knee push-up: men 50, women: 30. Do push-ups on the ground, but assuming a plank position on your knees.
- Incline push-up: men 50, women: 25. Do push-ups with your hands resting on a bench or other object elevated from the ground.
- Baseline push-ups: men 50, women 20
- Advanced variations: decline (legs elevated on a bench or chair), uneven hand positions, alternating wide/close hand positions, weighted vest push-ups.

Pull-ups: Keep your shoulder blades retracted during the pull to protect your spine. Keep your elbows towards your sides, lead with your chest up and keep your lower body quiet. Keep your chin slightly tucked to protect from cervical strain. Raise your chin over the bar and gradually lower all the way until your arms are straight. If you have elbow or shoulder issues, you can lower until just before your arms are straight to minimise joint strain.

Pull-up progressions: Baseline Essential Movement mastery: men 12, women 5.

- Chair-assisted: men 20, women 15. Start with your leg loosely positioned on a support chair underneath the bar. Engage your upper body muscles and use just enough leg force to assist getting your chin over the bar. You probably only need to use one leg but you can use two if necessary.
- Chin-up (inverted grip): men 7, women 4. Many find the chin-up to be slightly easier than a pull-up, particularly if you have any wrist, elbow or shoulder issues.
- Baseline pull-up (overhand grip): men 12, women 5
- Advanced pull-up variations: wide grip pull-up, uneven pull-up, weighted vest pull-up.

Squat: Stand with feet slightly wider than shoulder width and toes placed forward or turned slightly outwards in a natural position. Lower yourself by extending your backside out and bringing thighs to just below parallel to the ground. Stand back up completely, making sure your knees track in line with your feet.

Squat progressions: Baseline Essential Movement mastery: men 50, women 50.

- Assisted squat: men and women 50. Hold a pole or support object while lowering into and raising up from the squat position. Use the support object as little as possible.

- Baseline squat: men and women 50. Lower until thighs are parallel to ground, then stand up completely.
- Advanced squat variations: side-to-side squats (lowering on to one leg at a time), weighted vest squats.

Plank: Assume plank position by resting elbows on the ground, in alignment with shoulders, raising on to your toes and making your body horizontal from head to toe. Tuck your tailbone in a bit to alleviate potential strain on the lower back. Hold this position for the prescribed time period.

Plank progressions: Baseline Essential Movement mastery: men and women hold for 2 minutes.
- Forearm/knee plank: men and women 90 seconds. Assume plank position with your forearms and knees resting on ground. Tense your core and glutes during the exercise.
- Hand/feet plank: men and women 90 seconds. Assume plank position at the push-ups, with your hands and feet on the ground.
- Baseline forearm/feet plank: men and women 90 seconds. Some exercisers like to add a side plank to isolate on the lateral muscles of the abdominal wall. For a side plank, turn sideways, rest opposite hand or forearm on ground, and stack your feet on top of each other. Raise your hip so that your body is in straight head-to-toe alignment sideways and hold for 45 seconds on each side.
- Advanced plank variations: ups and downs (alternate between forearm/hand/forearm positions during count), one foot/one arm plank and side plank with raises, spidermans (drive right knee to right elbow from hand/feet plank position, return to plank position, repeat with left knee and elbow); cross spidermans (right knee touches left elbow and vice versa).

SPRINT SESSIONS

Sprinting emulates the true hunter-gatherer in all of us. These all-out efforts stimulate increased testosterone production, a slight pulse of human growth hormone (regulates muscle growth and recovery), an increase in muscle fibre growth and assorted other metabolic and fitness benefits. During the 21-day Challenge, two sprint sessions are suggested, with efforts scaled to your ability level to minimise risk of overdoing it. Sprint workouts can be running, if you can handle the impact, or using a bicycle or other cardio equipment for brief, all-out efforts with minimal or no impact.

An excellent basic sprint session can be conducted as follows: warm-up for at least five minutes with easy cardiovascular exercise. If you are running, I also recommend a series of dynamic stretches to really get the tendons and joints adapted to impending intense exercise. After your dynamic stretch sequence, initiate 4 to 6 'strides' where you gradually accelerate to your actual sprint speed for a few moments, then gradually ease back down into a jog. Each of these strides may take 15 seconds. Pay close attention to proper form during your strides, and take note of any pains or problem areas that might limit your efforts during the main workout set.

- Knee-to-chest: Start out gently pulling knees up to chest and releasing.
- Pull quads: Glasses off – time to get serious! Grab your foot from behind, pull gently up to your bottom and release.
- Open hips: Facing forward, rotate your knee up and along your bodyline, then place directly in front of you. This is especially important to improve hip flexor mobility that is compromised by sitting in a chair all day.

- Mini lunge: Take exaggerated-length steps, getting your front thigh to near parallel but not quite. Don't overdo this one, it's just a warm-up!
- Hopping drill: Warmup exercises progress to become more strenuous. Launch off one leg, driving knee high into chest, then land on the same leg. After a short hop forward, launch off opposite leg, driving knee high. Balance your launch effort between height and distance. Pump arms vigorously during each sequence.
- High knees: This one will get your heart rate up, and help you focus on achieving correct form during sprints. Run forward with exaggerated knee lift, striving to slap palms. During actual sprinting, focus on preserving a tall, straight body, driving knees high, and maintaining a balanced center of gravity through fast, efficient leg turnover.
- Grok Hang: Hang from a bar letting your body elongate naturally, and hang for as long as you can to obtain a full body stretch.
- Grok Squat: Lower in to a classic squat position, affording a gentle stretch for your lower legs, back and arms. Lower your torso in between your legs with your arms extended in front of you.

Once you are properly warmed up, initiate your main set of 6 sprints lasting 8–15 seconds at 80–90 per cent of maximum effort (effort range depending on your fitness and experience level). Accelerate gradually into each sprint to ease the risk of muscle strain. Take a 30–60 second recovery between each effort in order to return your breathing pattern to near normal. Your work and recovery times are also subject to your fitness level, so a novice might do six sprints of eight seconds, at 80 per cent effort, with a 60-second recovery interval between each sprint. An expert might do six sprints of 20 seconds, at 95–100 per cent effort, with a 30-second rest interval. Don't worry about your speed; the key here is to deliver a high-intensity effort. For less impact, conduct the identical set of 6 times 8–20 second efforts on a stationary bike or other cardio machine, or even sprinting uphill and jogging back downhill.

You may also wish to implement creative options such as plyometrics (explosive movements using maximum effort, such as repeated bounding, bunny hopping, stair climbing) into your sessions. Review the Sprint Workouts Appendix (page 264) for several more suggested workouts for different fitness levels. The key is to focus on the brief, intense, all-out aspect and refrain from a prolonged session that leads to exhaustion. Even accomplished athletes need not do more than 6 all-out sprints of 20 seconds, or 8–10 all-out sprints of 15 seconds. With sprinting, the focus is always on increasing your explosive speed, not doing more reps or going further.

> *The key with sprinting is to focus on brief, all-out efforts. Refrain from prolonged sessions that lead to exhaustion, and only sprint when you are 100 per cent rested and energised for a maximum effort.*

You should finish your sprint (and strength) training sessions with tired muscles, but feeling pleasantly invigorated and buzzed from the effort. Remember to always align your efforts with your daily levels of energy, motivation and health. Don't conduct a sprint workout just because your schedule suggests it is time. If you miss a planned workout or fall short of the ideal recommendations here, it's just not that big a deal. Your body will preserve fitness quite well even if your workout routine and total weekly volume ebbs and flows with the variables and challenges of daily life. In contrast, falling into a chronic exercise pattern will definitely compromise your fitness level and your health when you apply more stress than your body can handle without sufficient rest.

PLAY SESSIONS

I think my main motivation to work hard in the gym or at the park doing PEMs and sprints is for the pay-off I get when I play. My weekly pick-up game of Ultimate Frisbee is a pretty spirited competition. Every Sunday, a dozen or more of Malibu's finest gather at a city park and we go at it pretty hard for two hours. At 58, I'm assigned to match up 'mano-a-mano' with some impressive and much younger physical specimens, including high school varsity athletes such as my son Kyle. They seem to get faster each week! The work I put in during my PEM and sprint sessions helps me to play hard and to prevent injuries that might arise during my efforts to hang out with the younger generation.

During the 21-day Challenge, you will tackle three play assignments: a small play endeavour, a grand weekend endeavour, and figuring out ways to take mini play breaks during the day. Your formal play sessions ideally will involve being outdoors in sunlight, fresh air, open space and significant physical exertion. Even mini play breaks lasting 10 minutes can make a big difference. For example, stop off at the park on your way home from work and take a quick tour of the parcourse or playground equipment. Head over to the local courts and jump into a pick-up football game. Take your dog to the park and throw the frisbee, but surprise your pal by chasing him to the disc after you throw it!

ACTION ITEM 5:
SLOW LIFE DOWN

Going Primal involves reducing the complexity of your diet, exercise and lifestyle habits, and taking a simple, practical approach to concepts and challenges that we typically overanalyse. It's about finding the time and space to actually have fun and enjoy your life as it otherwise rushes by. Let's cover some specific ways to slow life down over the next 21 days in the areas of your diet, digital stimulation, exercise, personal time, relationships, sleeping habits and work.

DIET

If you can embrace the basic concepts of Primal eating and remember them for the rest of your life, you can reject the complexity and regimentation of Conventional Wisdom and cultivate a more relaxed and enjoyable relationship with food. During your 21-day Challenge, you will engage in various diet challenges that will connect you with the global 'slow food' movement, which promotes growing your own, buying locally, preparing meals from scratch, eating in a relaxed, celebratory environment and generally deepening your appreciation of the entire experience of eating.

Some important elements of the Primal eating style I'd like you to keep in mind immediately are to focus on getting maximum pleasure from your meals and engaging in relaxing mealtime habits. Eat all of your meals and snacks with full attention and awareness to the pleasure it provides you. Depart from regimented meals to the extent that your body actually experiences sensations of hunger, which will deepen your appreciation of the satisfaction food provides. Increase the frequency of IF efforts as you get more Primal adapted. Eat your food slowly and chew eat bite completely (20–30 chomps is ideal) to facilitate proper digestion. Avoid overeating by asking yourself, 'Am I really hungry for another bite, or have I had enough?' Paying attention to how food makes your body feel will naturally regulate your appetite and calorific intake.

To be clear, paying attention in this manner is the antithesis of the compulsive approach where portions, calories and food choices are obsessively regimented and measured against the arbitrary standards of some gimmicky diet. This is about paying attention to pleasure and satisfaction rather than a scorecard.

DIGITAL STIMULATION

Discipline your use of technology so you can leverage its advantages instead of becoming a slave to it. Here are a few tips for handling email, digital entertainment and hand-held devices. During your 21-day Challenge you will have lots of fun tackling these topics!

- **Email**: No one is arguing that going back to the days of licking stamps and sealing envelopes is preferred to the efficiency of email, but could email serve you even better if you used it in a more focused and disciplined manner? Your email challenge will involve spending specified time blocks to correspond by email, then shutting it down in between to focus on peak performance tasks. It's also critical to develop strategies to prioritise the communication you send and receive, paying more attention to high priority items and being disciplined to aggressively filter out low priority and unsolicited items.

- **Entertainment**: Use a DVR (Digital Video Recorder) to streamline your television watching experience, set your own entertainment schedule and eliminate exposure to commercials. Restrict the use of digital technology in the final two hours before bedtime. If you have kids, require that they turn in their digital technology (a big salad bowl on the kitchen table will do) after dark or until their homework is complete. Set aside time blocks to consume digital entertainment, social media, and news, and refrain from exposing yourself to a constant stream of distracting information titbits throughout the day.

- **Personal Digital Assistants (PDA):** Sure, it's okay to enjoy the flexibility and efficiency advantages of mobile technology. It beats being chained to a desk or repeatedly circling the school grounds looking for your kid after a Friday night football game. However, it's critical to ensure that your potential for hyper-connectivity does not interfere with cognitive down time, family time and meaningful interpersonal relationships. Establish 'business hours' for your PDA by shutting it off when engaged with family, recreation or when you've simply had enough. Realise that our genes are hard wired to be attuned to distraction, whether it be the vibration of a text message during the school play, or an ominous rustling in the bushes during Grok's time. There is simply no better solution than turning off the power so you are not even tempted to disengage from what you are doing at the moment.

EXERCISE

We have discussed at length the importance of rejecting Chronic Cardio in favour of aerobic sessions at moderate heart rates, along with increasing all forms of daily movement. It's also important to reject any feelings of compulsion, guilt or negativity about sub-par or missed workouts. Don't worry that slowing down your pace or missing workouts will cause you to get out of shape or gain weight (especially as you change from a sugar burner into fat-burning beast). Appreciate the process of getting fit, being outdoors, challenging your body to explore your physical limits and stop attaching your happiness and self-esteem to results.

If you are the competitive type, harness your instincts and grand ambitions so you can direct them in a productive manner on race day. Cultivate authentic and mutually supportive relationships with your training partners, instead of allowing every workout to become a surreptitious competition and ego-feeding frenzy. If your body feels sluggish on certain days and you fall short of predetermined goals, understand that fitness improvement comes through an optimal balance of stress and rest. A tired body or stiff, sore muscles are indicators that you are still in the process of getting fitter from previous workouts, provided you take time to recover! Here are a couple more tips to develop a more healthy and balanced approach to exercise:

- **Cultivate intuitive skills:** On a 1 to 10 scale, keep track of your energy level, motivation level and health/immune function each day, along with the degree of difficulty, performance and satisfaction level of your workouts. Try to get your workout scores aligned with your subjective scores, taking only what your body gives you each day and nothing more. For example, if you are having a rough day with scores of '4' in energy, motivation and health do a moderate workout that rates a '4' on a 1 to 10 difficulty scale.

  As you get into the rhythm of aligning workouts with energy, practise sitting quietly for a few moments before beginning your workouts and envisioning the session in your mind. Ask your-self if your planned session is really the right thing to do on that particular day, or envision a different session that might feel more aligned with your current physical and mental conditions. Don't be afraid to alter workouts midway through; pull the plug if your body is dragging, or push to a higher level if you are feel-ing fantastic.

- **Eliminate compulsive behaviours:** If you are a slave to your training log or have a tendency to follow a regimented schedule regardless of how you feel, force yourself to become more flex-ible. You could even go so far as throwing away your log – or at least put it aside for a month or two. Apply the same discipline and focus to resting as you do to training, so that you optimise both the stress and the rest sides of the balance scale. Remem-ber, Grok's coping strategy to survive in primal life was to do the absolute bare minimum necessary for survival. There were no extra gathering outings to pad out weekly mileage totals, nor killing of surplus animals to win shiny trophies.

PERSONAL TIME

Realise the need to regularly unplug from everything – digital stimu-lation, other people and the civilised world. Even if you only have 20, 10 or 5 minutes to spare on a particular day, take some time to sit and contemplate, take a stroll, birdwatch, and just slow down the pace of your thinking and moving to reflect and relax. Whether you are on a farm or in a high-rise flat, it's likely that you can find a quiet place to unplug and experience some solitude when you need it. Reject the flawed rat race notion that you are a bad mum or a corporate slacker if you take time out during a hectic day to decompress. Grab a few minutes of personal time each day (we will encourage your compliance during the 21-day Challenge), and also make an effort to carve out some prolonged sessions on weekends and holidays.

I call this taking my 'cave time'. Unfortunately, cave time frequently gets compromised when you need it most, especially when life gets busy and hectic. Often, my wife Carrie is there to provide a gentle reminder that my irritable, edgy behaviour might benefit from some cave time. I regularly use the time to take inventory and reflect upon things in my life that I am grateful for. After a brief self-imposed seclusion (usually a few hours of hiking), I return to my real world as good as new.

RELATIONSHIPS

The accelerated pace of technological progress is in discord with our primal genes. We are accustomed to less stimulation and a firm rooting in the present, physical world. Much has been written about the potential fall-out from digital life – kids missing out an a proper outdoor childhood of skinned knees and muddy clothes, burnt out, multitasking mums trying vainly to keep pace with cultural pressures, or harried executives chasing 'success' at the expense of health, family and sanity.

One element of slowing life down you will focus on during your 21-day Challenge is to de-emphasise your virtual relationships (email, texting, Facebook, Twitter, etc.) in favour of nurturing day to day interactions with family and friends. Envision your relationships in an 'intimate circle' of family and close friends, and a larger 'social circle' of co-workers, neighbours, casual friends, exercise partners and so forth. Sociologists believe that you are capable of maintaining a strong intimate circle of around 12 people, and a social circle of around 60 people. Attempting to maintain a larger social circle, through digital tools and other superficial communication, may compromise your ability to maintain a strong intimate circle, as well as a truly meaningful social circle. Sure, these numbers are merely theories, and it's difficult to draw demarcation lines at various friendship levels, but it can be valuable to absorb the spirit of this message and focus more on face time than Facebook.

- **Face time:** How about scheduling a lunch or a quick morning coffee with a business associate that you typically interact with online? It might be more productive than a hundred emails. The late Mark McCormack, old-time business guru, founder of the global sports conglomerate IMG, and author of *What They Don't Teach You At Harvard Business School*, related that a single round of golf with someone provided more character-revealing insights than prolonged traditional interactions in a business setting. Just be sure you aren't texting between holes!

- **Family-only time:** No digital stimulation or other distractions, just conversation, exercise/sports/leisure/play sessions, board games, art projects and other endeavours that you can do together and interact in a meaningful way.

- **Favour test:** Anthropologist and evolutionary biologist Robin Dunbar characterises an authentic and strong personal relationship by *an ability and willingness to do each other favours.* How many people would you bend over backwards for at a moment's notice? Would you lose a weekend helping them move across town, or rush to their aid in the event of a family emergency? Do a bit of analysis on how you spend your social time each day, and strive to put your family first, then nurture your social circle, and downscale your quest to build a bigger Twitter following. As Dunbar reminds us, 'A touch is worth a 1,000 words any day.'

> *Strong personal relationships are characterised by an ability and willingness to do each other favours. Strive to put family first, then your social circle, and back off on efforts to be a social media superstar.*

SLEEPING HABITS

For two million years, our circadian rhythms have been governed by the consistent rising and setting of the sun, a powerful natural phenomenon that has been artificially manipulated and widely disregarded only in the past century. Widespread sleep deprivation is one of the most destructive side effects of our fast paced, high-tech modern life. Excessive artificial light and digital stimulation after dark are the primary culprits in disturbing the flow of melatonin and other hormones that facilitate optimal sleep and restoration.

Soon after it gets dark, your genes are programmed to release melatonin into the bloodstream – a process known as Dim Light Melatonin Onset (DLMO). Melatonin causes you to feel drowsy by slowing down brain and metabolic functions and allowing you to gracefully switch from wake to sleep. Under ideal circumstances, the rising of the sun each morning, and your complete cycling through all phases of sleep, triggers a drop in melatonin and increased production of serotonin. Serotonin, aka the 'feel-good hormone', boosts your metabolic function, mood and energy levels, so you wake up feeling refreshed and ready for an active day.

> *Today, the sleep process is initiated when we make it dark, throwing us out of alignment with the sun and compromising our ability to fall asleep easily, sleep soundly and awaken refreshed.*

This elegant process has been all messed up by Thomas Edison, Bill Gates, Steve Jobs, TV and other facilitators of excess artificial light and digital stimulation after dark. Today, the sleep process is initiated when we *make it dark*, throwing us out of alignment with the sun and compromising our ability to fall asleep easily, sleep soundly and awaken refreshed. Follow these tips to slow down your evenings to gradually transition into a sleeping state and wake up refreshed each morning.

- **Dim the lights after dark:** Make a general effort to minimise illumination throughout your home after it gets dark. Utilise dimmer switches, install yellow-tinted light bulbs in frequently used lamps, light candles instead of flipping switches and wear a pair of yellow lens sunglasses or yellow safety glasses (cheapies can be found at DIY stores, or get a nice pair at smithoptics. eu). Using yellow bulbs, lenses and candles will help prevent artificial light from interfering with DLMO. Although it might seem a little goofy to don a pair of yellow lenses for your evening computer or TV time, these little touches can go a long way towards mitigating the damage produced by excessive artificial light in daily life.
- **Screen curfews:** If you insist on using computer, television or other screens after dark, do it earlier in the evening and devote the final two hours before bedtime to relaxing activities such as an evening stroll, quiet reading, conversation or family time (board games, etc.)
- **Get f.lux:** Install a free software program called f.lux on to computers you use at night-time (available at stereopsis.com). F.lux changes your screen display (technically the 'colour temperature'

of the screen) to align with ambient light, softening a light source as strong as the midday sun. Also, choose grey/minimal contrast backgrounds on software applications and browsers when possible.

HIGH-ENERGY MORNINGS

Here are a few tips to enjoy high-energy mornings. Even if you're not a morning person, we are all hard wired to respond to the rising of the great orb each day.

- **Get things right in the evening:** Allow DLMO to take effect so you get to bed on time and cycle through all the phases of sleep optimally.
- **Environment:** Keep your sleeping quarters relaxing, clutter-free, screen-free, cool 18–20°C (65–68°F), as dark as possible, and dedicated to rest or sleeping only. Awaken with sunlight or soothing, natural sounds instead of a buzzer.
- **Early to rise:** Awaken – hopefully naturally – as close as possible to sunrise. Expose yourself to bright light immediately upon awakening to kick-start serotonin production. Avoid disturbing alarms and try to use soothing, natural alarm functions if you must honour a designated wake-up time.
- **Gradually transition into a wakeful state:** Do some light yoga sequences, walk around the block with the dog, take a leisurely hot shower, read in bed for a few minutes or do some deep breathing in the garden. Skip the blaring morning news or moving at a hectic pace right away.

Getting adequate sleep and waking up refreshed and energised will improve your fat metabolism, eating habits/appetite regulation, workout performance, mood, energy levels and work productivity.

WORK

A great many job descriptions seem to demand (literally ... on the job description!) that you multitask, stay connected, respond quickly to various forms of communication, and generally do everything you can to speed the action up rather than slow down. Many experts are now second-guessing the ideal of a hyper-speed workplace. In Jim Collins's best-selling business strategy book *Good to Great*, he extols the 'hedgehog concept,' where individuals and entities that plod along with a passionate focus on a narrow area of expertise – honouring a 'single, organising idea' like a hedgehog – can prevail over 'foxes' who act more impetuously to compete incessantly in a broader economic realm. Case studies of companies that succeed with a deliberate, narrowly focused approach are offered, and contrasted against companies playing the fox that ultimately crash and burn (can you say, *'global economic collapse due to overextended financial institutions'*?) Collins himself takes about five years between books, but has produced a string of best-sellers.

Even ER nurses, who operate in an extremely high-stimulation crisis environment, where speed of care is a life or death matter, must remain calm, focused and centred amidst chaos. Ditto for athletes in the heat of battle. Legendary basketball coach John Wooden liked to convey this concept to his players with the quip, 'be quick, but don't hurry'. Sure, you have to keep pace with workplace norms and expectations, but do you really need to have the live chat window open while you are working on a new client proposal? Do you really need to text during your kid's football game or answer emails in the final hour of your night when you should be calmly transitioning into your evening sleep?

My productivity declines noticeably when my life or work habits fall out of balance. I might be racking up hours on the time clock, but being overstimulated, under-rested, undernourished or just plain burnt out plays out in the quality of work I produce – whether it be writing, compiling research in an organised manner or balancing my cheque book.

> *Don't succumb to the flawed Conventional Wisdom notion that productivity is directly related to time, without distinguishing quality from quantity. Take frequent breaks away from focused, sedentary efforts to refresh brain and body.*

Over the years, I have resolved to pay ever closer attention to my level of focus during the workday, and immediately take corrective action when lulls occur. This might mean stepping outside for a five-minute mini-workout, going down for a 20-minute nap, or even jumping in the car to run a couple of errands. The payoff is outstanding, but it requires trust and patience to back away when you need to. Don't succumb to the flawed Conventional Wisdom notion that productivity is directly related to time, without distinguishing quality from quantity. The following tips can help you reduce workplace stress and improve focus, productivity and job satisfaction.

- **Customise and prioritise information:** Too much of a good thing can spell trouble. Having the world's information at our fingertips helps our collective economic productivity skyrocket, but a constant assault of information in a variety of forms can overwhelm you to the point of permanent distractibility and diminished focus. Strictly filter and edit every bit of information you are exposed to during your workday, and in your whole life for that matter. Ignore or dispose of information that is unsolicited, irrelevant and uninteresting, such as roadside billboards, rubbish articles in the morning newspaper, television commercials, news broadcasts and Internet banner ads.

Remain proactive as much as possible throughout the day, focusing on the highest priority task (with a little help from your boss perhaps!), and refuse to be pulled away by potential distractors. Learn to say 'no' when you need to, and negotiate with yourself and others to carve out the time and personal space you require to achieve maximum productivity and fulfilment in your work.

- **Focus on a single peak performance task at a time:** The vaunted ability to multitask in the modern workplace is a big fat ruse: the human brain is literally incapable of multitasking. Instead, when faced with two or more sources of stimulation at the same time, we divert our attention back and forth in quick spurts, compromising our ability to focus on any single task. Sure, you can probably rake leaves and chat with a friend on your Bluetooth with no ill effects, but when complexity escalates (e.g. negotiating an important business deal on the phone while driving in rush hour traffic), your attention span, creativity and intellectual ability will decline, your stress will increase and you will get that familiar frazzled feeling that epitomises rat race-induced burnout.

  Reject the entire concept of multi-tasking in favour of focusing. Develop the discipline to ruthlessly eliminate all forms of distraction when you engage in a peak performance task. Honour the concept even with mundane tasks (don't talk and drive; don't even talk and rake!) to hone your focusing skills into a habit.

- **Take frequent breaks:** Take a 1 to 3 minute break for every 20 minutes of intent focus at work, particularly if you are sedentary. Stand up, walk around a bit, refocus your eyes on distant objects, do some simple stretches or strengthening exercises, or simply close your eyes and chill for a while before turning your attention back to the screen or other focused engagement. If you've been busy moving around in a noisy warehouse, retail sales floor or emergency room, you might prefer to go sit in your car and listen to some classical music.

  Take a 10-minute break every 2 hours, and a midday break of 30 minutes or more. Get up, get moving and change your venue. Take a stroll in the office courtyard if you've been seated at your desk. Find a small private space and haul off a few PEM exercises, or sprint up a few flights of stairs at the office building.

THE 21-DAY CHALLENGE

TIME TO GET TO IT!

INSTRUCTIONS

The following section offers a reasonably-paced schedule of challenges in the categories of diet, exercise and lifestyle (usually one challenge per day in each category), for 21 days. Tackling these challenges will give you the knowledge and practical experience to custom design an enjoyable, flexible Primal lifestyle over the long term. At no time should you feel overburdened, overstressed or exhausted by these challenges. For the most part, they require minimal time, and are appropriate to all levels of fitness and previous knowledge about diet, exercise and health.

Start the 21-day Challenge when your energy and enthusiasm are high, and you have a minimal amount of distractions and overall life stress. If the present time is inconvenient, establish a suitable future date (after a holiday, a half-year term or fiscal quarter). Please start Day 1 on a Monday, since certain endeavours are designed with a workday or weekend in mind. Each day includes some brief journal exercises and a blank to assign an overall daily 'success score' for how well you think you completed the challenges. Rank your success scores on a scale of 1–10, with 10 being the best.

If you anticipate some resistance from your family, friends, roommates or doctor, I suggest you gently inform them that you are implementing some new health practices and would appreciate their support. Try to keep the drama to a minimum and avoid knockdown, drag-out debates about grains, cholesterol or contrary fitness principles. In the months ahead, walking your talk with improved mood, energy, attitude and body composition will be far more powerful than any grand proclamations you make at the outset.

DAY 1

- **Diet – kitchen purge:** First things first! Time to overhaul your kitchen and eliminate typical Western diet foods that may be lurking in your house. Refer to Action Item 1 (see page 116) to complete a successful purge. It should only take an hour to give you some breathing room to go Primal. You may feel more trepidation over this action item than any other, for these foods heading to the landfill have provided you with a convenient source of quick energy and immediate gratification your entire life (don't worry, Primal shopping sprees are imminent!)

  There is no halfway allowed here. A purge is a purge, and even those precious old favourites must go if you want to transform from a sugar burner into a fat-burning beast in a relatively short time period.

- **Diet – restock preparation:** Grab a few essentials at your local supermarket so you don't starve before tomorrow: eggs, veggies, fruit, nuts and meat. Meanwhile, spend a little time on the telephone or Internet lining up the best locations for a proper Primal shopping spree tomorrow. Get ready to storm your local farmers' market or health food shop. If your local area lacks great options, refer to the Internet suggestions mentioned in Action Item 2 (see page 121).

- **Exercise – increase daily movement:** Think of ways to add more basic movement into your daily routine. Can you park purposefully far away at your office and at stores, avoid elevators, and bike or walk to errands instead of drive? Refer to the suggestions in Action Item 4 (see page 160) and implement at least two basic movement suggestions today – a morning walk with the dog, brief work breaks, a stroll before arriving home or an evening stroll with the family after dinner.

- **Lifestyle – Primal essentials:** Here are some basics you'll need to complete your 21-Day Challenge. Take inventory today and grab anything that's missing:
  - Evening tools: f.lux software on computer, DVR to optimise screen time, yellow lenses/lightbulbs and candles (see page 183) to minimise light.
  - Exercise basics: Comfortable clothing, suitable location for aerobic workouts, PEMs, and sprinting (back garden, health club, park or school). Pull-up bar for PEM. Weighted vest, if desired, to increase PEM degree of difficulty.
  - Kitchen basics: Fridge, freezer, cookbooks, cooking ware, spices.
  - Minimalist footwear: Not mandatory, but a popular element of the Primal movement.
  - Shopping: Internet access for online nutrient calculations and for discovering food shopping options.
  - Sleeping: Dark, cool, quiet, uncluttered sleeping environment.
  - Workplace: Find a convenient location at work for quiet reflective time and also quick exercise breaks. Consider adapting your computer desk for periods of stand-up work.

DAY 1 JOURNAL

Success score (1–10): _____

Kitchen Purge

Hardest part: _____

Best part: _____

Comments: _____

Restock Preparation

Primal shopping resources: _____

Comments: _____

Increase Daily Movement

Ways you can increase daily movement over the long term:

1: _____

2: _____

3: _____

4: _____

5: _____

Describe today's movement endeavours (morning, midday, evening stroll, etc.):

1: _____

2: _____

3: _____

Primal Essentials

Items acquired today: _____

Comments: _____

Summary Comments:

Daily energy levels 1–10: _____

Hunger level between meals 1–10: _____

Satisfaction level with meals 1–10: _____

Struggles today with Primal efforts: _____

Benefits noticed from Primal efforts: _____

Daily highlight(s): _____

Daily needs to improve: _____

DAY 2

- **Diet – Primal shopping spree:** Here is a shopping list for Primal cooking essentials, breakfast omelette, lunch salad, meat and vegetable dinner, and Primal snacks. Refer to Action Items 2 and 3 (see pages 121 and 129) for more details about choosing wisely in each food category.

  - Bacon: Some supermarkets offer convenient ready-cooked crispy smoked bacon. Microwave for 30 seconds and drape over your Primal omelette!

  - Cooking Essentials: Coconut oil, milk and flakes. Organic butter, extra-virgin olive oil.

  - Cheese: Grated mozzarella or Cheddar to sprinkle on to your omelette.

  - Eggs: Grass-fed or certified organic. Grab a couple of dozen and make a 3 to 4 egg Primal omelette for your breakfast. It's a great way to stabilise blood glucose and stay satisfied for hours.

  - Flavourings/Sauces/Condiments: Capers, coconut milk, Dijon mustard, fresh herbs, garlic, ginger, lemons, limes, nut butters, olives, pesto, sauerkraut, shallots, sun-dried tomatoes, tamari, unsweetened coconut flakes, vinegars.

- Fruit: If berries and stoned fruits are in season, grab some great local selections. Exercise moderation in this category if it's off season or you want to quickly shed excess body fat.

- Herbs and Spices: Hopefully your spice rack is already stocked with the basics: bay leaves, black pepper, chilli powder, cinnamon, cumin, garlic powder, nutmeg, onion powder, oregano, chilli flakes, sea salt. To expand your flavour horizons add allspice, cardamom, coriander, fennel seeds, paprika and turmeric. Spice blends such as curry powder, seasoned pepper and the like are an easy way to add a lot of flavour at once.

- Meat, Fish, Poultry: Use the highest quality and most appealing dinner main meals you can find, hopefully local/ pasture-raised/grass-fed or organic meat and sustainable fish. Major supermarkets have lots of organic meat, poultry and sustainable fish on offer.

- Nuts: Macadamia nuts are the best option, or use other favourites in snacks and salads.

- Snacks: Choose your favourites from the list on page 250.

- Vegetables – Omelette: Peppers, onions, mushrooms and spinach.

- Vegetables – Salad: Avocado, cabbage, carrots, cherry tomatoes, green or red onion, cos lettuce head or packet of baby salad leaves.

- Vegetables – Dinner: Beetroot, broccoli, cabbage, aubergine, leafy greens (chard, kale, pak choi), squash and courgettes are great to steam or stir-fry.

- **Exercise – moderate duration aerobic workout:** Determine 'moderate duration' according to your typical workout routine, likely somewhere in the range of 20–60 minutes. Maintain a comfortable pace of 55–75 per cent of maximum heart rate, tracked by wireless heart rate monitor or a few pulse checks during the session.
- **Lifestyle – calm, relaxing evening:** Establish a screen curfew after dark, and dim the lights and/or use yellow lenses and bulbs after dark. Take a 5–15-minute evening stroll or enjoy some family relaxation time (board game, cards, talking, reading). Awaken early, hopefully near sunrise and without an alarm, but do the best you can. Expose yourself to direct sunlight as soon as you wake up, and consider an energising morning ritual (breathing and stretching exercises, cold water plunge, hot shower, neighbourhood stroll), if you are not the morning type.

DAY 2 JOURNAL

Success score (1–10): _____

Primal Shopping Spree

Stores visited: _____

Hardest part: _____

Best part: _____

Comments: _____

_____

Moderate Duration Aerobic Workout

Location: _____

Activity: _____ Duration: _____

Comments: _____
_____

Calm, Relaxing Evening
Suggestions you followed: _____
Hardest part: _____
Best part: _____
Comments: _____
_____

Summary Comments:
Daily energy levels 1–10: _____
Hunger level between meals 1–10: _____
Satisfaction level with meals 1–10: _____
Struggles today with Primal efforts: _____
Benefits noticed from Primal efforts: _____
Daily highlight(s): _____
Daily needs to improve: _____

DAY 3

- **Diet – Primal celebration dinner planning**: Extend invitations to friends and loved ones for an authentic home-cooked meal on Day 6 or 7. Or, consider a dinner party where guests agree to bring their own Primal approved, homemade offerings. Take your time planning a creative menu and compiling a shopping list for recipe ingredients. Purchase the foods at a convenient time over the next couple of days. If you don't feel like entertaining, celebrate with yourself, significant other or family.

- **Diet – Boycott industrialised food:** Try to completely avoid eating at fast food chains or any other processed or frozen meals from multinational food corporations – today and for the duration of your 21-day Challenge.

  Today, make a specific statement in favour of your health and against the industrialisation of food by planting a few seeds in your garden, finding healthy local or mail-order alternatives to mainstream sources, or perhaps selling off any shares you have in Kraft or McDonald's!

- **Exercise – Full-length PEM workout:** Five minute warm-up of easy cardiovascular exercise, and two sets of maximum repetitions of push-ups, squats, pull-ups and abdominal planks.

- **Exercise/lifestyle – play:** Take a spontaneous play break for at least 20 minutes. Refer to the suggestions on page 173. Make plans for a grand play outing on Day 6 – invite family/friends, schedule lessons/car rentals and make all possible advance preparations for a weekend adventure.

DAY 3 JOURNAL
Success score (1–10): _____

Primal Celebration Dinner Planning
Menu choice: _____
Guests invited: _____
Comments: _____
_____

Boycott Industrialised Food
Steps taken against industrialisation of food: _____

Hardest part: _____

Best part: _____

Comments: _____

_____

Full-length PEM Workout

Success score: _____

Location: _____ Duration: _____

Reps completed:

Push-ups set 1: _____ set 2: _____

Squats set 1: _____ set 2: _____

Pull-ups set 1: _____ set 2: _____

Plank (time) set 1: _____ set 2: _____

Comments: _____

_____

Play

Spontaneous play session: _____

Location: _____ Duration: _____

Hardest part: _____

Best part: _____

Comments: _____

_____

Steps taken towards grand play outing: _____

Hardest part: _____

Best part: _____

Comments: _____

_____

Summary Comments:

Daily energy levels 1–10: _____

Hunger level between meals 1–10: _____

Satisfaction level with meals 1–10: _____

Struggles today with Primal efforts: _____

Benefits noticed from Primal efforts: _____

Daily highlight(s): _____

Daily needs to improve: _____

DAY 4

- **Diet – honour hunger:** Instead of eating in a regimented pattern guided by the clock, allow hunger sensations to guide your eating habits today. See how long you can last in the morning until you detect actual hunger sensations kicking in, or until energy levels dwindle slightly. Then, enjoy a delicious Primal meal, eating enough to feel satisfied but not full. Avoid overeating by asking yourself, 'Am I really hungry for another bite, or have I had enough?'

  Go about your day until you again notice hunger sensations, cravings or diminished energy requiring calories. Enjoy a delicious Primal lunch or snack, then repeat the process in the evening. This exercise will help strengthen your intuitive eating habits and prepare you for the IF challenges later on, when you are more Primal adapted.

- **Exercise – aerobic adventure:** Conduct an extended duration aerobic workout at 55–75 per cent of maximum heart rate. Try something new, such as rent a stand-up paddle board, a pair of snowshoes, inline skates or a mountain bike. Be safe, but step outside of your comfort zone, especially if most of your aerobic exercise is on gym machines.

- **Lifestyle – stand-up work station:** If you are an office worker or otherwise work for long periods sitting down, modify your work environment so that you can engage in periods of standing up. Grab a few reams of paper, some file boxes or anything else handy to lift your keyboard and monitor. If you work on a large corporate campus, you may be able to ask the Human Resources (HR) department to install a high shelf in your cube or office.

Try to stand up for as long as comfortably possible before sitting down. Repeat standing up as many times as you can after sufficient rest periods. Try to remove your shoes while you stand to get a barefoot experience.

DAY 4 JOURNAL
Success score: _____

Honour Hunger
What times did you eat?: _____

Hardest part: _____

Best part: _____

Comments: _____

_____

Aerobic Adventure
Location: _____

Activity: _____ Duration: _____

Comments: _____

_____

Stand-up Work Station

Modifications made: _____

Total time standing up: _____

Longest single stretch: _____

Hardest part: _____

Best part: _____

Comments: _____

_____

Summary Comments:

Daily energy levels 1–10: _____

Hunger level between meals 1–10: _____

Satisfaction level with meals 1–10: _____

Struggles today with Primal efforts: _____

Benefits noticed from Primal efforts: _____

Daily highlight(s): _____

Daily needs to improve: _____

DAY 5

- **Diet – modern foraging:** Dine out for a meal or two and see how well you can stay Primal aligned. Sharpen your assertiveness skills if you have to negotiate with the restaurant to alter their menu.

- **Exercise – abbreviated PEM workout:** Do one set of maximum reps for each of the four PEMs. After a five-minute warm-up of easy cardio, carry on until your muscles fail on each exercise, and take enough rest between exercises to return your breathing to normal.

- **Lifestyle – active couch potato rebellion:** Don't go longer than 30 minutes today without taking a movement break of 2 to 5 minutes. If you endure a long commute, exit the motorway halfway through and romp around in a park or field for a few minutes. Even if you are watching a good DVD, you can break into doing some plank and squat sets to honour this challenge.

DAY 5 JOURNAL

Success score: _____

Modern Foraging

Location: _____

Meal: _____

Location: _____

Meal: _____

Hardest part: _____

Best part: _____

Comments: _____

_____

Abbreviated PEM Workout

Success score: _____

Location: _____ Duration: _____

Reps completed:

Push-ups: _____

Squats: _____

Pull-ups: _____

Plank (time): _____

Comments: _____

Active Couch Potato Rebellion

Activity: _____ Duration: _____

Activity: _____ Duration: _____

Activity: _____ Duration: _____

Longest sedentary period without break today: _____

Hardest part: _____

Best part: _____

Comments: _____

_____

Summary Comments:

Daily energy levels 1–10: _____

Hunger level between meals 1–10: _____

Satisfaction level with meals 1–10: _____

Struggles today with Primal efforts: _____

Benefits noticed from Primal efforts: _____

Daily highlight(s): _____

Daily needs to improve: _____

DAY 6

- **Diet – Primal celebration dinner:** Shop for ingredients, prepare recipes and enjoy your celebration. Savour every bite and establish a tradition of social gatherings and delicious food. After the meal, if weather permits, encourage everyone to take a stroll outdoors for at least 10 minutes. A leisurely evening walk promotes efficient digestion, helps relax the mind and body after a busy day and facilitates an easy transition into a good night's sleep later in the evening.

- **Exercise – extended duration aerobic workout:** Conduct an extended duration aerobic workout at 55–75 per cent of maximum heart rate, lasting at least an hour and up to several hours, if you have the fitness base.

- **Lifestyle – listening challenge:** At your dinner party, conversation will likely drift to the impetus for the event and your efforts to go Primal. Challenge yourself to be an exceptional listener this evening. Instead of commanding the floor and providing a blow-by-blow account of your exciting journey, continually direct the attention and dialogue back to others. Focus on gathering information and insights that may help you become an effective guide and mentor for those who may some day aspire to their own 21-day Transformation. Notice how the energy and attention subtly shifts from one person to another during a gathering, and do your best to support and encourage others to communicate freely.

DAILY 6 JOURNAL

Success score: _____

Primal Celebration Dinner

Guests: _____

Menu: _____

Hardest part: _____

Best part: _____

Comments: _____

_____

Extended Duration Aerobic Workout

Location: _____

Activity: _____ Duration: _____

Comments: _____

_____

Listening Challenge

Hardest part: _____

Best part: _____

Did increased awareness improve your listening skills?: _____

Were you able to support and encourage others to communicate

freely? _____

Comments: _____

_____

Summary Comments:

Daily energy levels 1–10: _____

Hunger level between meals 1–10: _____

Satisfaction level with meals 1–10: _____

Struggles today with Primal efforts: _____

Benefits noticed from Primal efforts: _____

Daily highlight(s): _____

Daily needs to improve: _____

DAY 7

- **Diet – intuitive meals:** On the heels of your well-planned dinner celebration, anything goes today. Eat whatever foods you feel like eating without regard to cultural breakfast, lunch or dinner traditions. Perhaps you'll want last night's leftovers for breakfast? An omelette for dinner? Dark chocolate and macadamia nuts for a midday snack?

If you feel compelled to indulge in a non-Primal approved old favourite, go ahead and do so. Pay close attention to how your comfort food makes your body feel, including any unpleasant after-effects. See if you notice any emotional influence in your eating habits that might be compromising your health. For every bite you take throughout the day, focus on the enjoyment of the experience. Let go of any feelings of guilt, anxiety or other negative emotions associated with eating. From now on, your job is to attain maximum pleasure from your food choices each day.

- **Exercise/lifestyle – grand play outing:** Kayak, rock climb, cycle, picnic, hike, play Ultimate Frisbee, horseshoe, football, basketball or anything else that's fun, active and celebrates the great outdoors with family and friends!

DAY 7 JOURNAL
Success score: _____

Intuitive Meals
Hardest part: _____
Best part: _____
Comments: _____
_____

Grand Play Outing
Location: _____

Activity: _____ Duration: _____
Comments: _____
_____

Summary Comments:

Daily energy levels 1–10: _____

Hunger level between meals 1–10: _____

Satisfaction level with meals 1–10: _____

Struggles today with Primal efforts: _____

Benefits noticed from Primal efforts: _____

Daily highlight(s): _____

Daily needs to improve: _____

Week 1 Reflections

Week 1 success score: _____

Diet success score: _____

Comments on diet challenges: _____

Exercise success score: _____

Comments on exercise challenges: _____

Lifestyle success score: _____

Comments on lifestyle challenges: _____

Weekly highlight(s): _____

Weekly needs to improve: _____

What specific steps can you take to address your needs to improve

list? _____

Overall comments on week 1: _____

_____

DAY 8

- **Diet – no labels challenge:** Conventional Wisdom suggests you scrutinise the government-mandated 'Nutrition Facts' labels on all packaged foods – watching your fat grams, Recommended Daily Allowance (RDA) percentages or sugar/carbohydrate ratios. As you learnt in the Key Concepts, going Primal transcends most of this sugar-burner advice.

  Today your challenge is to give Nutrition Facts the boot by avoiding any foods that have a label on them! Emphasise local produce, farm fresh animal products, homemade nut butter or jerky, and other non-industrialised options. Take the challenge as far as you like. While no one will scorn you for eating a carrot, see if you can find fresh carrots over sliced, washed and bagged options, and pasture-raised eggs from a local farmer over a commercially produced carton of eggs.

- **Exercise – sprint workout:** Scale your effort to your experience with sprinting. If this is your first attempt, try a no- or low-impact exercise and make your hard efforts about 80 per cent of maximum. If you have a decent level of comfort and experience, sprint at 90 per cent of maximum effort. Be sure to warm up and cool down effectively, and keep the emphasis on quality instead of quantity.

- **Lifestyle – tiptoe into the barefoot world:** Try to go barefoot or use minimalist footwear for at least 60 minutes today. Let your feet breathe and become re-engaged in the act of walking, jogging and supporting your standing weight. If you have no experience, start gradually by simply walking about the house barefoot. When you become comfortable, you can introduce brief bouts of barefoot workout time.

DAY 8 JOURNAL:

Success score: _____

No Labels Challenge

Hardest part: _____

Best part: _____

Comments: _____

_____

Sprint Workout

Success score: _____

Location: _____

Activity: _____ Total duration: _____

Reps: _____ Duration or distance: _____ Rest interval: _____

Comments: _____

_____

Tiptoe Into The Barefoot World

Total time spent barefoot or in minimalist footwear: _____

Longest single stretch going barefoot or in minimalist footwear
(count standing or moving only, not sitting): _____

Hardest part: _____

Best part: _____

Comments: _____

_____

Summary Comments:

Daily energy levels 1–10: _____

Hunger level between meals 1–10: _____

Satisfaction level with meals 1–10: _____

Struggles today with Primal efforts: _____

Benefits noticed from Primal efforts: _____

Daily highlight(s): _____

Daily needs to improve: _____

DAY 9

- **Diet – track macronutrient intake:** It may be helpful to occasionally complete an exercise where you track your calorific and macronutrient intake through an online calculator. Most importantly, you must obtain adequate protein calories to preserve or build muscle tissue, and keep your carbohydrate intake in line with your body composition goals: 50–100 grams per day for fat loss, and 100–150 grams to maintain ideal body composition.

  Proceed with this exercise as follows: Write down everything you eat for an entire day. Use measuring tools (measuring jug, tablespoon, scales) to obtain accurate quantities. Carry around a small notepad so you don't forget anything. Visit fitday.com (my favourite of several Internet options), create a free account and begin inputting your foods into their database. After inputting all of your foods, Fitday will produce a simple pie chart with tabulations for daily protein, carbs, fat and total calories. Your data will be saved in their data base and you can repeat the exercise whenever you want to generate more daily reports.

- **Exercise – increase daily movement, part 2:** Hopefully you've been making a concerted effort to leverage the challenges from Day 1 (increase daily movement), Day 2 (stand-up work station) and Day 5 (movement breaks every 30 minutes) to discover lots of ways to move more throughout the day. Today I'd like you to redouble your efforts to park purposefully far away, avoid lifts and escalators, walk or cycle instead of drive, go no longer than 30 minutes without a brief movement break, stand instead of sit, take mini play breaks and implement any other creative movement ideas into your routine.

- **Lifestyle – media fast:** Ingest all of your news and information today during 10-minute time blocks in the morning and in the evening. Take this challenge seriously and try – just for a single day – to refrain from exposing yourself to a constant stream of distracting information and entertainment titbits throughout the day. During your 10-minute sessions, be proactive by choosing your favourite media resources, scanning headlines quickly and reading only stories that really interest you.

  This challenge may reveal how distracting a typical day might be when you fail to discipline yourself against constant and overwhelming digital stimulation. Hopefully you will build some awareness, battle against overstimulation and sharpen your focus on peak performance tasks.

DAY 9 JOURNAL

Success score: _____

Track Macronutrient Intake

Daily protein grams: _____

Daily carb grams: _____

Daily fat grams: _____

Total calories: _____

Areas to improve: _____

Worst part of results: _____

Best part: _____

Comments: _____

_____

Increase Daily Movement, Part 2

Ways you increased movement today:

1: _____

2: _____

3: _____

4: _____

Hardest part: _____

Best part: _____

Comments: _____

_____

Media Fast

Resources used (paper, website, TV, etc.) _____

Comments on morning block: _____

Comments on evening block: _____

Comments on media fast period: _____

Hardest part: _____

Best part: _____

Comments: _____

_____

Summary Comments:

Daily energy levels 1–10: _____

Hunger level between meals 1–10: _____

Satisfaction level with meals 1–10: _____

Struggles today with Primal efforts: _____

Benefits noticed from Primal efforts: _____

Daily highlight(s): _____

Daily needs to improve: _____

DAY 10

- **Diet – eating environment**: Today you will focus on creating an optimal eating environment for all meals. No more wolfing food on the go – even if it's Primal aligned! Take the time to create attractive place settings and calm, quiet environments dedicated to eating only. Eliminate all distractions such as computers, television and high-energy music. Refrain from reading a newspaper or magazines while eating. Instead, focus on quiet conversation with your eating companion(s), or self-reflection if alone.

  Make a concerted effort to slow the pace of your eating. Chew eat bite completely to facilitate good digestion and maximum satisfaction. Taking 20–30 bites will enable the enzymes in your saliva to lubricate food for easy transportation through the oesophagus, and break down macronutrients for efficient digestion in the stomach.

- **Exercise – Primal WOW**: WOW is 'Workout of the Week', a popular theme at www.marksdailyapple.com where a creative workout honouring the Primal Blueprint Fitness philosophy is suggested each week. Today you will combine the PEM exercises with some sprinting and plyometrics to produce the following fun and challenging session. The distances and rest intervals suggested are chosen according to your fitness level.

* **Push-ups** (or appropriate push-up progression exercise): one set maximum reps
* **Sprint**: immediately sprint 40–80 metres
* **Rest:** 30–60 seconds
* **Pull-ups** (or appropriate progression exercise): one set maximum reps
* **Bunny hops**: hop (feet together for take-off and landing) for 20–40 metres
* **Rest:** 30–60 seconds
* **Squats** (or appropriate squat progression exercise): one set maximum reps
* **Sprint**: immediately sprint 40–80 metres
* **Plank**: one set maximum time

Advanced Additions

* **Sprint/bunny hop:** After planks, immediately sprint 80 metres, hop 40 metres in opposite direction, then walk 40 metres to starting point. Rest 30 seconds, then repeat.

* **Lifestyle – email fast:** Engage in email correspondence only during specified morning and afternoon time periods of 30 minutes or less. Between these efforts, disengage from your email server and pursue focused, productive tasks without succumbing to distraction. Even a heavy hitter should be able to completely organise an inbox and reply to key messages in less than 30 minutes.

Prioritise your emails as follows: first, filter your inbox by mass deleting any correspondence that was not solicited and/ or not important. Next, overview your remaining messages to identify urgent and/or highest priority messages. A quick look at the sender, subject line or first few lines of text in the preview mode will help achieve this quickly. Use software tools (star, red flag, etc.) to highlight these messages if desired, then proceed through them in approximate order of priority. Finally, handle low priority, but necessary, correspondence with succinct replies and clearly defined responses such as 'yes', 'no, sorry', or 'I'll have that for you by Friday'. This will help maintain positive working and social relationships without draining excessive time from higher priority tasks.

DAILY 10 JOURNAL

Success score: _____

Eating Environment

Comments on setting: _____

Comments on distractions: _____

Comments on pace: _____

Hardest part: _____

Best part: _____

Comments: _____

_____

Primal WOW

Success score: _____

Location: _____ Duration: _____

Running distances (40 or 80m): _____

Push-ups: _____

Pull-ups: _____

Squats: _____

Plank (time): _____

Advanced options: _____

Email Fast

Comments on morning correspondence period: _____

Comments on afternoon correspondence period: _____

Suggestions implemented: _____

Comments on fasting period: _____

Was your communication more efficient during correspondence period knowing your time was limited? _____

Did you notice any increased productivity during fasting period?:

_____

Any decreased productivity during fasting period?: _____

Hardest part: _____

Best part: _____

How can you integrate the email fast concept into your daily workday? _____

Overall comments: _____

_____

Summary Comments:

Daily energy levels 1–10: _____

Hunger level between meals 1–10: _____

Satisfaction level with meals 1–10: _____

Struggles today with Primal efforts: _____

Benefits noticed from Primal efforts: _____

Daily highlight(s): _____

Daily needs to improve: _____

DAY 11

- **Diet – create a Primal recipe:** Concoct your own Primal main meal, writing down all the ingredients and measuring portions for future reference. If the first go is not quite right, revise the recipe and repeat the exercise tomorrow.
- **Exercise – moderate duration aerobic workout:** Go 20–60 minutes at 55–75 per cent of maximum heart rate.
- **Lifestyle – work peak performance**: Start your workday by carefully creating a to-do list with tasks ranked in order of priority. Methodically proceed through each task with undivided attention. If a phone call brings an urgent request from your boss requiring an hour of work, revise your to-do list accordingly and stick to the renegotiated plan.

  During this exercise, you may notice how easily you are pulled away from your priorities by unrelenting stimulation. Resolve to honour your to-do list for the duration of the day. While it does take a little extra diligence and preparation time to be proactive instead of reactive at work, building these skills can dramatically reduce stress and increase productivity.

DAILY 11 JOURNAL

Success score: _____

Create a Primal Recipe

Recipe ingredients: _____

Recipe instructions: _____

Hardest part: _____

Best part: _____

Proposed revisions (if necessary): _____

Comments: _____

_____

Moderate Duration Aerobic Workout

Activity: _____ Duration: _____

Location: _____

Comments: _____

_____

Work Peak Performance

Increased productivity/reduced stress from to-do list exercise:

_____

Drawbacks/increased stress from to-do list exercise: _____

Hardest part: _____

Best part: _____

Elements you can integrate into typical workday: _____

Overall comments: _____

_____

Summary Comments:

Daily energy levels 1–10: _____

Hunger level between meals 1–10: _____

Satisfaction level with meals 1–10: _____

Struggles today with Primal efforts: _____

Benefits noticed from Primal efforts: _____

Daily highlight(s): _____

Daily needs to improve:_____

DAY 12

- **Diet – Primal recipe, part 2:** Repeat the recipe from yesterday if you feel it requires improvement.

- **Diet – go coco-nuts:** Purchase an arsenal of coconut products and prepare a coconut-focused recipe or two. Coconut oil can replace PUFA oils (see page 15) for cooking. Coconut flour is an excellent substitute for any flour recipe, including pancakes. Coconut milk is a delicious milk substitute, great to pour on berries, drink straight, blend into a smoothie or whisk into a whipped cream. Coconut flakes or chips are versatile recipe and meal accompaniments. All can be found at a decent health food shop or in major supermarkets.

  For recipes, a Primal Fuel Smoothie (see page 256) is an easy shortcut here (since coconut milk solids form its base). The *Primal Blueprint Cookbook* and *Primal Blueprint Quick and Easy Meals* have many coconut options, and entering 'coconut' into the search bar at marksdailyapple.com will produce a handful of excellent posts with recipes.

- **Exercise – PEM workout:** Choose an abbreviated or full session depending on your energy and motivation level today. Go until your muscles fail on each exercise, and take enough rest between exercises to return your breathing to normal.

- **Lifestyle – nurture your intimate and social circles:** Schedule a family social gathering with no digital stimulation or distractions. Enjoy conversation, exercise/sports/leisure/play sessions, board games, art projects and other endeavours that you can do together. Schedule a lunch or quick morning coffee with a business associate whom you typically interact with via digital communication. Finally, disconnect from all social media and strictly limit consumption of digital media as you did on Day 9 to allow for more meaningful and prolonged social interaction.

DAY 12 JOURNAL

Success score: _____

Primal Recipe, Part 2 (if applicable): _____

Changes made: _____

Comments: _____

_____

Go Coco-nuts

Products used: _____

Elements you will integrate into diet long term: _____

Hardest part: _____

Best part: _____

Comments: _____

_____

PEM Workout

Success score: _____

Location: _____ Duration: _____

Reps completed:

Push-ups set 1: _____   set 2: _____

Squats set 1: _____   set 2: _____

Pull-ups set 1: _____   set 2: _____

Plank (time) set 1: _____   set 2: _____

Comments: _____

_____

Nurture Your Intimate and Social Circles

Intimate circle connections: _____

Social circle connections: _____

Hardest part: _____

Best part: _____

Elements to integrate into lifestyle long term: _____

Comments: _____

_____

Summary Comments:

Daily energy levels 1–10: _____

Hunger level between meals 1–10: _____

Satisfaction level with meals 1–10: _____

Struggles today with Primal efforts: _____

Benefits noticed from Primal efforts: _____

Daily highlight(s): _____

Daily needs to improve: _____

DAY 13

- **Diet – top of spectrum:** Today your challenge is to acquire a food ranked at the top of the spectrum in each food category as follows:
    - Local, grass-fed meat or poultry. Failing that, find some certified organic meat.
    - Local, grass-fed eggs. Failing that, find some certified organic eggs.
    - Sustainable fresh fish or oily, cold-water canned fish.

- Locally grown produce. Failing that, find some certified organic produce grown domestically.
- Macadamia nuts – raw or dry-roasted.
- Organically grown extra-virgin, first cold-press olive oil.
- Raw, fermented, unpasteurised, grass-fed dairy products (cheese, butter, cream cheese, ghee, kefir, whole milk, yogurt).
- Indulgences: If you have the need to indulge, find some high-quality red wine and/or dark chocolate at 75 per cent cocoa content or higher.

- **Exercise – moderate duration aerobic workout:** Conduct a moderate duration aerobic workout at 55–75 per cent of maximum heart rate. Explore a new exercise today, or repeat an activity you enjoyed from your aerobic adventure day.
- **Lifestyle – get adequate sunlight:** This challenge is ideal for the summer months with relatively strong sunlight. If this is impossible to tackle today, read through the challenge and try it when the weather allows.

Vitamin D plays a critical role in regulating healthy cellular function, yet widespread deficiency occurs in the developed world due to sedentary lifestyles and irrational fears of skin cancer. Failure to obtain the bare minimum of sun exposure and vitamin D production actually increases your risk for a variety of cancers, including melanoma! While this is a complex subject warranting deeper discussion, becoming familiar with the basics of safe, adequate sun exposure is a valuable component of your 21-day Challenge.

While individual circumstances vary wildly (pigment, latitude, time of day/year, weather, ozone, altitude, surface reflectiveness), maintaining a slight tan year-round indicates that your vitamin

D blood levels are adequate. When tanning is a challenge (e.g. winter), taking vitamin D supplements can help boost your levels into the target range. Contrary to Conventional Wisdom, getting adequate vitamin D from dietary sources alone is virtually impossible. Vitamin D experts recommend you obtain around 4,000 International Units (IU) per day, but the typical Western diet provides only around 300 IU per day, and the vaunted glass of milk provides only around 100 IU per day. By contrast, 20–40 minutes of direct sunlight can produce around 10,000 IU of vitamin D, which can easily be stored in your cells for future use.

For maximum vitamin D production, expose large surface areas of your skin (arms, legs, torso) to direct sun for about *half of the amount of time it takes to get slightly burnt.* Your personal 'time to burning' on a particular day is an estimate dependent upon the aforementioned variables, but trying to hit the half-way mark is a pretty low-risk endeavour. Today, try to expose as much skin as possible to the sun for a reasonable length of time, avoiding burning of course. To reduce skin cancer concerns, cover frequently exposed, sensitive skin areas (face, neck, hands) with clothing or sunscreen.

DAY 13 JOURNAL

Success score: _____

Top of Spectrum

Products acquired:

Meat: _____   Location: _____

Eggs: _____   Location: _____

Fish: _____   Location: _____

Macadamia nuts: _____ Location: _____

Extra-virgin olive oil: _____ Location: _____

Dairy products: _____ Location: _____

Indulgences: _____ Location: _____

Describe the budget impact: _____

Hardest part: _____

Best part: _____

Purchases you can sustain long term: _____

Purchases unrealistic to sustain long term: _____

Next best choice to sustain long term: _____

Comments: _____

_____

Moderate Duration Aerobic Workout

Activity: _____ Duration: _____

Location: _____

Comments: _____

_____

Get Adequate Sunlight

Was today an appropriate day to get some sun exposure?: _____

Minutes of exposure: _____ Parts of body: _____

Hardest part: _____

Best part: _____

Suggestions you can implement long term: _____

Summary Comments:

Daily energy levels 1–10: _____

Hunger level between meals 1–10: _____

Satisfaction level with meals 1–10: _____

Struggles today with Primal efforts: _____

Benefits noticed from Primal efforts: _____

Daily highlight(s): _____

Daily needs to improve: _____

DAY 14

- **Diet – sweet spot**: Whip out your notepad and do another fitday.com macronutrient analysis of everything you eat, paying particular attention to hitting the Primal Blueprint Carbohydrate Curve Sweet Spot (see page 67) of between 50 and 100 grams of total carbohydrate intake. Accept this challenge even if you don't have fat reduction goals, but refrain from doing any extreme workouts.

- **Diet – IF Alert:** Tomorrow your challenge will be to IF from tonight's dinner to as long as you can comfortably last without eating tomorrow. The essence of the exercise is to check how Primal adapted you have become by following up a day of carb restriction with an IF. Honour the spirit of the challenges and don't react to this advance warning by pigging out tonight!

- **Exercise – rest day:** Take a break from exercise today, and reflect upon how planned rest periods – even if you are feeling great – are an essential element of an effective exercise programme. Harness your energy for some challenging sessions coming up!

- **Lifestyle – reflection:** Take some time to comment in detail about your second week of Primal challenges.

DAY 14 JOURNAL

Success score: _____

Sweet Spot

Daily protein grams: _____

Daily carb grams: _____

Daily fat grams: _____

Total calories: _____

Areas to improve: _____

Worst part of results: _____

Best part: _____

Comments: _____

_____

Rest Day

Comments: _____

_____

Summary Comments:

Daily energy levels 1–10: _____

Hunger level between meals 1–10: _____

Satisfaction level with meals 1–10: _____

Struggles today with Primal efforts: _____

Benefits noticed from Primal efforts: _____

Daily highlight(s): _____

Daily needs to improve: _____

Week 2 Reflections

Week 2 success score: _____

Days 1 to 14 success score: _____

Discuss how you addressed your needs to improve list from week 1: _____

Diet success score: _____

Comments on diet challenges: _____

Exercise success score: _____

Comments on exercise challenges: _____

Lifestyle success score: _____

Comments on lifestyle challenges: _____

Weekly highlight(s): _____

Weekly needs to improve: _____

What specific steps can you take to address your needs to improve list? _____

Overall comments on week 2: _____

_____

DAY 15

- **Diet – IF 1:** You should be reasonably Primal adapted at this point and able to succeed with this challenge at some level. You shouldn't be famished upon awakening, particularly since yesterday's challenge was to minimise your carb intake/insulin production. Your fast will run from last night's dinner for as long as you can last today without calories. When you experience strong sensations of hunger or diminished energy levels, enjoy a delicious and satisfying Primal meal.

- **Exercise – Mark's favourite Primal workout:** Ready for another exciting high-intensity challenge? Here's one of my custom designed favourites. See how you like it! Novice exercisers choose the shorter distance and easier PEM; vice-versa for advanced exercisers. Really advanced exercisers can tackle this with a weighted vest if they dare … Take 30–60 seconds' rest between exercises to catch your breath. Refer to the detailed descriptions of these exercises on pages indicated.

1. **Back slide arches (page 168)/spidermans (page 170):** 25 or 50 metres (82 or 164 feet), down with arches, back with spidermans. Take a 15-second rest before return trip.

2. **Push-ups (page 168):** One set, maximum reps. Advanced exercisers can do decline push-ups.

3. **Bunny hop (page 266)/lunge walk:** 25 or 50 metres (82 or 164 feet), down with bunny hops, back with lunge walks. Take 30-seconds' rest before the return trip.

4. **Pull-ups (page 168):** One set, maximum reps.

5. **Stairs or hopping drill (pages 172, 265):** Attack the stadium or jump for joy!

6. **Plank (page 170):** Maximum time to failure at appropriate plank progression exercise.

7. **Sprint (page 171):** 2 x 50 or 75 metres (82 or 246 feet). Take a 15-second rest before return trip.

8. **Squat (page 169):** One set, maximum reps. Why did I put this last? Oh man, good luck with that!

- **Lifestyle – cave time:** Take 30–60 minutes of solo time today where you disengage from digital stimulation, other people and all other influences of the civilised world. I find a walk to be the best option, but meditating on a park bench is just fine too. Get some space, slow down your thoughts and just relax and reflect.

DAY 15 JOURNAL

Success score: _____

IF 1

Duration: _____ Start time: _____ End time: _____

Hardest part: _____

Best part: _____

Comments: _____

_____

Mark's Favourite Primal Workout

Success score: _____

Distance chosen for arch/spider and bunny/lunge: _____

Push-up style: _____ Reps: _____

Pull-up reps: _____

Stairs or skipping: _____ Describe: _____

Plank time: _____

Sprint distance chosen: _____

Squat reps: _____

Comments: _____

_____

Cave Time

Duration: _____

Where did you go? _____

Hardest part: _____

Best part: _____

Comments: _____

_____

Summary Comments:

Daily energy levels 1–10: _____

Hunger level between meals 1–10: _____

Satisfaction level with meals 1–10: _____

Struggles today with Primal efforts: _____

Benefits noticed from Primal efforts: _____

Daily highlight(s): _____

Daily needs to improve: _____

DAY 16

- **Diet – go local**: Strive to obtain half or more of your calories from local sources today. Hopefully you can take advantage of a farmers' market to enjoy some great produce and animal products over the next few days.

- **Exercise – moderate duration aerobic workout:** Exercise for 20–60 minutes at 55–75 per cent of maximum heart rate. Note that the previous 16 days have involved a pretty ambitious Primal exercise regimen. If you have been significantly exceeding your normal level of exercise, take the rest you need to produce a strong finish over the final five days. If this means skipping today's workout and hitting the minimum recommended range for future aerobic workouts, then that's fine.

- **Lifestyle – calm, relaxing evening 2**: Up the ante from your Day 2 challenge by making an even more dramatic effort to have a mellow evening. Completely refrain from screen time this evening in favour of a family board game or reading. Linger at the dinner table for conversation in true European style instead of rushing off to digital entertainment. Take an extended neighbourhood stroll of 15–30 minutes to enjoy some fresh air, open space and casual conversation or self-reflection. Get by on a minimum of artificial light after the sun sets, and get to sleep early in a dark, quiet, relaxing room.

Tomorrow, wake up as close as possible to sunrise and immediately expose yourself to direct sunlight. Choose an energising morning ritual: breathing and stretching exercises, a brief neighbourhood stroll, cold water plunge or your Day 17 abbreviated PEM workout if you are so inclined.

DAY 16 JOURNAL

Success score: _____

Go Local

Estimate percentage of calories obtained locally: _____

Places you shopped: _____

Hardest part: _____

Best part: _____

Comments: _____

_____

Moderate Duration Aerobic Workout

Activity: _____ Duration: _____

Location: _____

Comments: _____

_____

Calm, Relaxing Evening 2

Suggestions implemented: _____

Hardest part: _____

Best part: _____

Comments: _____

Summary Comments:

Daily energy levels 1–10: _____

Hunger level between meals 1–10: _____

Satisfaction level with meals 1–10: _____

Struggles today with Primal efforts: _____

Benefits noticed from Primal efforts: _____

Daily highlight(s): _____

Daily needs to improve: _____

DAY 17

- **Diet – modern foraging 2:** Escalate your challenge from Day 5 by eating all of your meals outside of the home and staying Primal aligned. Seek out a new restaurant or market and navigate the offerings to create Primal meals.
- **Exercise – full-length PEM workout:** One set, maximum reps of the four PEMs. Go until your muscles completely fail on each exercise, and take enough rest between exercises to return your breathing to normal. If you are feeling a bit tired or sore, reduce this session to an abbreviated PEM, and/or complete 75 per cent of your estimated reps until failure (e.g. do 15 push-ups if 20 is failure).

- **Lifestyle – reach out:** Discover an open-minded and deserving family member or friend who might be receptive to going Primal and initiate some dialogue about how you might help him or her. Engaging with someone about lifestyle change is a sensitive issue and must be handled accordingly. The operative word here is 'receptive', which is easy to determine in a brief conversation.

    If you approach this challenge with an intervention vibe, you are likely to be rebuffed. Instead, take a casual, positive approach by inviting your friend to join you for a PEM workout, or a Primal meal at your home. You can even get them a copy of this book with an enthusiastic recommendation and invitation to discuss further. Offer information, guidance and camaraderie on demand so your friend can control the dynamics of this journey. And hey, if no one comes to mind immediately, don't force it. Keep this challenge in the back of your mind and try it when the time is right.

DAY 17 JOURNAL

Success score: _____

Modern Foraging 2

Location: _____ Meal: _____

Location: _____ Meal: _____

Location: _____ Meal: _____

Hardest part: _____

Best part: _____

Comments: _____

_____

Full-length PEM Workout

Success score: _____

Location: _____ Duration: _____

Reps completed:

Push-ups set 1: _____   set 2: _____

Squats set 1: _____   set 2: _____

Pull-ups set 1: _____   set 2: _____

Plank (time) set 1: _____   set 2: _____

Comments: _____

_____

Reach Out

Who did you talk to?: _____

Best part: _____

Most difficult part: _____

Comments: _____

_____

Summary Comments:

Daily energy levels 1–10: _____

Hunger level between meals 1–10: _____

Satisfaction level with meals 1–10: _____

Struggles today with Primal efforts: _____

Benefits noticed from Primal efforts: _____

Daily highlight(s): _____

Daily needs to improve: _____

DAY 18

- **Diet – kitchen purge 2:** Take a final sweep of the premises and get rid of any non-Primal offenders still hanging around. If you are juggling the interests of a non-Primal significant other, kid(s) or flatmate(s), see if you can arrange to segregate Primal and non-Primal items to specific shelves in order to sharpen your focus.

- **Exercise – moderate duration aerobic workout:** Exercise for 20–60 minutes at 55–75 per cent of maximum heart rate. Rest or hit the minimum if you are a bit tired or sore.

- **Lifestyle – use your brain:** Today you will take the initial steps to pursue a creative intellectual challenge that provides a refreshing break and balance to the responsibilities of hectic daily life. Sign up for lessons in a foreign language, a musical instrument or dancing. Start a jigsaw or crossword puzzle right now, write a short story or initiate a DIY or landscaping project. Try anything else outside of your comfort zone that sounds interesting and challenging!

  Also, figure out ways throughout the day to keep your brain challenged and stimulated. Replay a song on your iPod to memorise the lyrics, find your school yearbook and try to recall the names of long-lost fellow students or use your head instead of a calculator to add up numbers.

DAY 18 JOURNAL

Success score: _____

Kitchen Purge 2

Offending foods tossed: _____

Compromises with cohabitants: _____

Hardest part: _____

Best part: _____

Comments: _____

_____

Moderate Duration Aerobic Workout

Activity: _____ Duration: _____

Location: _____

Comments: _____

_____

Use Your Brain

Daily challenge: _____

Steps taken for long-term challenge: _____

Best part: _____

Most difficult part: _____

Comments: _____

_____

Summary Comments:

Daily energy levels 1–10: _____

Hunger level between meals 1–10: _____

Satisfaction level with meals 1–10: _____

Struggles today with Primal efforts: _____

Benefits noticed from Primal efforts: _____

Daily highlight(s): _____

Daily needs to improve: _____

DAY 19

- **Diet – go local 2**: Strive to obtain an estimated two-thirds of your calories from local sources. Shop at the right places, eliminate processed products and try to identify the source of everything that enters your mouth today. Keep an eye towards local in the future and see if you can maintain a high percentage of calorific intake from local sources.

- **Exercise – sprint workout:** Step up the intensity to between 90–100 per cent of maximum effort. Go all-out if you have the experience and are structurally sound enough to handle it. Otherwise, refer to the Sprint Workout Suggestions Appendix (page 264) to pick a suitable running or low-impact workout option.

- **Lifestyle – mini play breaks:** Take three spontaneous play breaks lasting 5–15 minutes today. Get up from your desk or out of your car and exercise your free spirit. Find a kid, animal or co-worker and toss a ball, kick a can or climb a fence. Look around and notice how possibilities abound – even in the drabbest of city parks or the most crowded airports!

DAY 19 JOURNAL

Success score: _____

Go Local 2

Estimated percentage of calories obtained locally: _____

Places you shopped: _____

Hardest part: _____

Best part: _____

Comments: _____

_____

Sprint Workout

Success score: _____

Location: _____

Activity: _____ Total duration: _____

Reps: _____ Duration or distance: _____ Rest interval: _____

Comments: _____

_____

Mini Play Breaks

Break 1: _____ Location: _____ Duration: _____

Break 2: _____ Location: _____ Duration: _____

Break 3: _____ Location: _____ Duration: _____

Hardest part: _____

Best part: _____

Comments: _____

_____

Summary Comments:

Daily energy levels 1–10: _____

Hunger level between meals 1–10: _____

Satisfaction level with meals 1–10: _____

Struggles today with Primal efforts: _____

Benefits noticed from Primal efforts: _____

Daily highlight(s): _____

Daily needs to improve: _____

DAY 20

- **Diet – IF 2:** Repeat the challenge and see if you can last any longer. Remember, this is not a pressure-packed suffer-fest, but rather a simple intuitive exercise to resist eating until you are actually hungry.
- **Exercise – extended duration aerobic workout:** Conduct an extended duration aerobic workout at 55–75 per cent of maximum heart rate, lasting at least an hour, and up to several hours if you have the fitness base.
- **Lifestyle – 10 Primal changes:** Compile a list of things you do today that represent a distinct transition from your pre-Primal lifestyle behaviours. For example, awakening early and getting some sunlight, eating a low insulin-producing breakfast, taking a break for movement or play, standing up while working, disciplining your use of email, moderating heart rate during an aerobic workout, dimming the lights and donning yellow lenses after dark and so on.

   Can you get the list up to 10? How about 15? 20? Enjoy the challenge and see if you can appreciate the accumulating benefits of transforming to a Primal lifestyle.

DAY 20 JOURNAL

Success score: _____

IF 2

Duration: _____ Start time:_____ End time: _____

Hardest part: _____

Best part: _____

Comments: _____

_____

Extended Duration Aerobic Workout

Activity: _____ Duration: _____

Location: _____

Comments: _____

_____

Primal Changes (minimum of 10, maximum of)

1: _____

2: _____

3: _____

4: _____

5: _____

6: _____

7: _____

8: _____

9: _____

10: _____

11: _____

12: _____

13: _____

14: _____

15: _____

Hardest part: _____

Best part: _____

Easiest to sustain long term: _____

Most difficult to sustain long term: _____

Comments: _____

_____

Summary Comments:

Daily energy levels 1–10: _____

Hunger level between meals 1–10: _____

Satisfaction level with meals 1–10: _____

Struggles today with Primal efforts: _____

Benefits noticed from Primal efforts: _____

Daily highlight(s): _____

Daily needs to improve: _____

DAY 21

- **Diet – sensible indulgences:** Make a list of non-Primal foods, drinks or habits that have been the most difficult to manage during your challenge. Still have a hankering for that morning croissant at the coffee house, or a late-night spoonful (or 2 or 12) of ice cream? Heighten your awareness of any shortcomings, accept them without judgement or negativity, and formulate a plan to make your indulgences more sensible.

  If you are a chocoholic, can you swap milk chocolate products for dark chocolate? Can the rich, satisfying taste of a Primal Fuel smoothie (see page 256) take the place of a coffee house muffin when you need a quick morning snack on the go? Can beef jerky and apple slices take the place of a commercial energy bar when an afternoon lull hits? Can you smuggle a bag of macadamia nuts into the cinema to replace absent-minded scoffing of popcorn? If these replacement suggestions don't fly, can you in fact appreciate a slice of cheesecake once a month without feeling guilty and diving into a five-day sugar binge? Maintain a positive mindset, focus on pleasure and satisfaction and see if you can find ways to tweak your indulgent habits without feeling deprived or frustrated.

- **Exercise – full-length PEM workout:** Complete a full-length PEM workout consisting of a five-minute warm-up period of easy cardiovascular exercise, and two sets of maximum repetitions of push-ups, squats, pull-ups and planks.

- **Lifestyle – after photo:** If you are inclined, snap a progress photo to compare to the one you may have taken at the outset of the 21-day Challenge. If you have been diligent in completing the challenges and aligned with Primal eating, you should notice some improvements in body composition.

    Hopefully you will feel that the 21-day Transformation Action Items are sustainable over the coming months and years, and have confidence that you can continue to progress at a reasonable rate with body composition goals until you attain your ideal. Take a photo every 60 days to record your progress.

- **Lifestyle – reflection:** Take some time to comment in detail about your third week of Primal challenges, and your entire 21-day journey.

DAY 21 JOURNAL
Success score: _____

Sensible Indulgences
Difficult to eliminate: _____

Potential Primal swap: _____

Difficult to eliminate: _____

Potential Primal swap: _____

Difficult to eliminate: _____

Potential Primal swap: _____

Full-length PEM Workout

Success score: _____

Location: _____ Duration: _____

Reps completed:

Push-ups set 1: _____   set 2: _____

Squats set 1: _____   set 2: _____

Pull-ups set 1: _____   set 2: _____

Plank (time) set 1: _____   set 2: _____

Comments: _____

_____

'After' Photo

Comments: _____

_____

Summary Comments:

Daily energy levels 1–10: _____

Hunger level between meals 1–10: _____

Satisfaction level with meals 1–10: _____

Struggles today with Primal efforts: _____

Benefits noticed from Primal efforts: _____

Daily highlight(s): _____

Daily needs to improve: _____

Week 3 Reflections

Week 3 success score: _____

Discuss how you addressed your needs to improve list from week 2:

_____

_____

_____

Diet success score: _____

Comments on diet challenges: _____

Exercise success score: _____

Comments on exercise challenges: _____

Lifestyle success score: _____

Comments on lifestyle challenges: _____

Weekly highlight(s): _____

Weekly needs to improve: _____

What specific steps can you take to address your needs to improve list? _____

_____

Overall comments on week 3: _____

_____

_____

_____

21-day Transformation Summary

Overall diet success score: _____

Overall exercise success score: _____

Overall lifestyle success score: _____

Hardest parts: _____

_____

_____

Best parts: _____

_____

_____

Remaining needs to improve items: _____

_____

_____

Overall comments on 21-day challenge: _____

_____

_____

_____

APPENDIX

THE PRIMAL FOOD SPECTRUM AT A GLANCE

MEAT AND POULTRY
- Local, pastured animals or 100 per cent grass-fed animals
- FSA-certified organic animals
- Minimise CAFO (concentrated animal feeding operation) meats
- Avoid heavily processed meat (hot dogs, continental sausage, salami, etc.)

EGGS
- Local, pastured eggs
- FSA-certified organic eggs
- Conventional eggs
- Avoid egg white mixtures lacking the most nutritious part – the yolk!

FISH
- Wild-caught, oily, cold-water fish from remote, pollution-free waters (high omega-3)
- Troll or pole line-caught (instead of commercial longline); lower on food chain (less toxin concern); ocean-friendly harvesting (study rankings from aforementioned websites)
- Approved farmed fish (salmon, shellfish, trout, catfish, crayfish, tilapia)
- Avoid: other farmed fish (toxin exposure, diminished nutritional value); Asian imports (polluted waters, lax regulations)

VEGETABLES

- Locally grown, pesticide-free, in-season vegetables
- Choose organic for edible skins/large surface area (leafy greens, peppers) to minimise pesticide risk
- Minimise or avoid conventionally grown, remote, out-of-season vegetables

FRUIT

- Locally grown, pesticide-free, in-season fruits
- Emphasise high-antioxidant/low-glycaemic (berries, stone fruits)
- Choose organic for edible skins/difficult washing to minimise pesticide risk (berries)
- Moderate intake to support fat-reduction efforts – respect Total Metabolic Fructose scores
- Avoid remote, conventionally grown, out-of-season fruits

MACADAMIA NUTS

High in monounsaturated fat, antioxidants and essential aminos. Better omega-6:omega-3 ratio than other nuts. Best snack choice

FATS AND OILS

- Best food sources: Avocados, macadamia nuts, omega-3 fish, saturated animal fats (pastured/grass-fed), coconut products
- Extra-virgin olive oil: Domestically grown, first cold-press only
- Bottled or capsuled omega-3 oil supplements
- Cook with coconut oil, butter, or other saturated animal fat
- Eliminate trans and partially hydrogenated fats and PUFA oils

MODERATION FOODS

- High-fat dairy: Emphasise raw, fermented, unpasteurised products (aged cheese, butter, cream, cream cheese, ghee, kefir, whole milk, yogurt)
- Other nuts, seeds and derivative butters: High nutritional value, but exercise some moderation to promote dietary omega-6:omega-3 balance
- Supplemental carbohydrates: Sweet potatoes and yams, quinoa and wild rice for hard-training athletes or other high-calorie burners without excess body fat concerns
- Beverages: Water, mineral water, carbonated water, tea, coffee

HERBS AND SPICES

Excellent antioxidant value and enhanced flavouring. Organic, dried herbs preferred.

SENSIBLE INDULGENCES

- Alcohol: Red wine is best (high-antioxidant), beer is okay. Drink moderately and responsibly. 'First to burn' calories can compromise weight-loss efforts.
- Dark chocolate: High antioxidant value, brain-stimulating agents, euphoric agents and satisfying saturated fat. Choose highest possible cocoa content (75 per cent or higher)

SNACKS

- Berries
- Dark chocolate
- Fish: Canned herring, mackerel, sardines, tuna
- Hard-boiled eggs

- Jerky: Homemade or otherwise minimally processed
- Macadamia nuts: Probably the ultimate healthy, incredibly satisfying between-meal snack
- Olives: A great source of monounsaturated fatty acids
- Vegetables with nut butter spread

FOODS TO AVOID - AT A GLANCE

- **Beverages**: Designer coffees; energy and sports drinks; bottled, fresh and powdered juices; flavoured milks; soft drinks (including diet); sweetened coffees, cocktails, powdered mixes and teas
- **Baking ingredients**: Flours, starches, sweeteners, syrups, yeast
- **Condiments**: Jams, jellies, ketchup, mayo, low-fat dressings and other products with PUFA or HFCS
- **Dairy**: Frozen yogurt; ice cream; low-fat and nonfat milks and yogurts; processed cheeses
- **Fats and oils**: PUFA oils (canola, buttery spreads and sprays, vegetable oils, margarine, shortening); trans and partially hydrogenated oils
- **Fast food**: Boycott industrialised food – try Intermittent Fasting instead!
- **Fish**: Farmed (with certain exceptions); objectionable catching methods/polluted waters; Asian imports; top of food chain species
- **Grains**: Cereal, corn, pasta, rice, wheat, and derivative products: bread/flour/crackers, breakfast foods, crisps, cooking grains, puffed snacks
- **Pulses**: Alfalfa, beans, lentils, peanuts, peanut butter, soya beans, tofu
- **Meat**: Pre-packaged/heavily processed, smoked, cured, nitrate-treated meats. Limit CAFO meats in favour of pastured/grass-fed
- **Processed foods**: Energy bars; frozen breakfast, dinner and dessert products; fruit bars and rolls; granola bars; high-protein bars; packaged, grain and sugar-laden snack products

- **Sweets**: Candy; cake; cookies; doughnuts; ice cream; milk chocolate; pie; ice lollies and other frozen desserts; all forms of sugar and sweeteners; syrups, other packaged/processed sweets and treats

PRIMAL ESSENTIAL MEALS

The *Primal Blueprint Cookbook* and the *Primal Blueprint Quick and Easy Meals* offer over 100 recipes in each book, in various meal, snack and even dessert categories. Even if you are not the gourmet type, you can discover some excellent meals that are easy to prepare even when you are pressed for time. Following is a brief description of some simple, quick Primal Essential Meals that I enjoy on a regular basis. So, you have your PEM workouts and now you have your PEM meals! Remember, personal preference and flexibility are hallmarks of the Primal eating style. Feel free to experiment with foods and recipes that please you the most. Don't be afraid to simplify or modify recipes, or repeat your favourite meals over and over.

BREAKFAST – PRIMAL OMELETTE

While constructing a beautiful omelette can be well worth the effort, breakfast doesn't need to be any more difficult than frying up a few eggs. Seriously, if you aren't one to gather ingredients and follow a step-by-step recipe, just do this: Melt butter in a pan, add eggs, remove when cooked. Done! It doesn't need to be any fancier than that. If you want a real treat that will kick-start your day and optimise fat metabolism, give this omelette and its many amazing variations a try:

3–5 eggs, grass-fed or organic
Butter for pan-frying
56–113 g (2–4 oz) chopped spinach, mushrooms, onions, peppers,
  tomatoes
Salt, pepper, mixed seasonings as desired

- **Preparation:** Sauté the chopped vegetables in a separate pan with an ample amount of butter. Set aside, prepared for quick inclusion into the omelette. Whisk the eggs until smooth and pour into a buttered frying pan over a medium heat. Allow the eggs to begin cooking, then use a spatula to carefully scrape the egg away from the pan edge as it sets, allowing any uncooked egg to flow into the scraped-away space. Lightly shaking the pan helps the eggs to set evenly. When the eggs are set (from liquid into solid), carefully spread your vegetables over one half of the pan. With the spatula, carefully release and then lift the entire empty side of the omelette and fold on to the ingredients. Lightly press down on the omelette to seal. Cover the pan with lid for additional cooking if the omelette appears runny inside. Scoop or slide the omelette out carefully and serve.

- **Primal Omelette Variations:**
  Avocado: Garnish finished omelette with slices of avocado.
  Cheese: Sprinkle liberal amounts of Cheddar, mozzarella or feta cheese into centre of omelette before folding.
  Meat: Add pieces of chopped bacon, chicken or steak to centre of omelette. Bacon and cheese can replace the vegetables if desired.
  Other ways with eggs: Prepare eggs baked (frittata), hard-boiled, fried, poached or scrambled. Mix in vegetables and meat options or serve as side dishes.

Spice variations and other additions: Try salsa, soured cream and avocado for Mexican style. Try green and red chilli powder, cumin and coriander leaves for Indian style. Try pesto or marinara sauce with extra cheese for Italian style.

BREAKFAST – PRIMAL FUEL SMOOTHIE

This Primal Nutrition meal replacement product is available at primalblueprint.com.

500 ml (18 fl oz)) liquid (water or coconut milk)
150 g (5 oz) ice
2–3 scoops of Primal Fuel

- **Preparation:** Process all ingredients thoroughly in a blender. Gradually add water, stopping the blender to stir the ingredients if you are having trouble, until the mixture is completely blended.

- **Primal Fuel Smoothie Variations:**
  Coconut flakes: Provides even more coconut flavour and texture for true aficionados.
  Fruit: Frozen, peeled bananas or frozen berries add new flavours to keep things interesting.
  Nut butter: Macadamia, almond or cashew butter go well with both chocolate and vanilla Primal Fuel.

MORE BREAKFAST IDEAS

- Cottage cheese (full-fat) with berries and nut butter
- Egg muffins (search 'omelet muffins' on marksdailyapple.com)
- Leftover dinner from the night before

- Smoked salmon with cream cheese on cucumber slices
- Macadamia nuts and berries – grab a handful of each and go!
- Meat (steak, bacon, sausage) and greens (sautéed or raw)
- 'Noatmeal' (search 'oat-free oatmeal' on marksdailyapple.com)

- **From the *Primal Blueprint Quick & Easy Meals* Cookbook:** Berry pancake; Breakfast burrito; Broccoli quiche

LUNCH – PRIMAL SALAD

I don't call this a 'two-minute big-ass salad' for nothing! You don't really need a recipe for this one; just throw whatever salad leaves, raw veggies and cooked meat you have on hand into a big bowl, dress it with olive oil and lemon, sprinkle on some nuts or seeds, and – *voilà* – your Primal salad is complete. If you want to pack as much flavour as possible into the bowl, check out the salad variations that ensure you'll never grow tired of this fast and healthy lunchtime staple.

Large handful of salad leaves, such as baby salad, cos or spinach
Large handful of chopped protein, such as cooked steak, chicken, turkey or wild salmon
Small handfuls of various vegetables, aiming for 4 to 6 different types (e.g. peppers, carrots, celery, cherry tomatoes, cucumbers, onions, radishes)
Small handful of nuts and/or seeds

Dressing: *113 g (4 oz) extra-virgin olive oil*
57 g (2 oz) freshly squeezed lemon juice
½ teaspoon sea salt
¼ teaspoon freshly ground black pepper
15 g (½ oz) finely chopped fresh herbs of choice

- **Preparation:** Ideally, keep containers of chopped vegetables in your fridge to quickly access all week. The same goes for protein – always cook extra meat at dinner to use for lunch the next day. This way, you can make a Primal salad in around two minutes. Toss salad ingredients in a large bowl. Whisk together dressing ingredients in a separate bowl (or put into a jar and shake well). You will have plenty left over for future salads. Liberally cover the salad with dressing and enjoy. If you are taking your salad to work, use a large plastic container and keep the dressing separate until it's time to eat.

- **Primal Salad Variations:**
  Dressing: Use the dressing recipe above but instead of olive oil try sesame, avocado, macadamia or walnut oil. Experiment with different types of vinegar instead of lemon juice. Whisk coconut milk with curry powder or other herbs and spices for a quick creamy dressing.
  Garnishes: A sprinkling of berries, dried fruit, pumpkin seeds or sunflower seeds, olives or chopped bacon adds more flavour and texture.
  Greens: Branch out beyond plain old lettuce for added flavour and nutrition. Try kale, purslane, mache, lamb's lettuce, rocket, cabbage and seaweed.
  Meat: Canned tuna, salmon, sardines, herring, anchovies and smoked salmon are a fast way to add protein.
  Toppings: Avocado, hard-boiled egg and cheese add satiating fat.

LUNCH – PRIMAL WRAP

Large leaf of iceberg lettuce for the wrap
Favourite sandwich ingredients (see Primal Wrap Variations below)

- **Preparation:** Place the lettuce leaf on a plate and top with your desired sauces or spreads. Layer the fillings, putting the heaviest ingredients, such as meat, in first and stopping short of the edges of the leaf to prevent any spillages. Try to fold the edges in to form a barrier, then carefully roll the leaf around the fillings.

- **Primal Wrap Variations:**
  Carne asada: Skirt steak, hot chillies, onions, salsa, spices, (cayenne pepper, oregano, salt, thyme) and tomatoes.
  Chicken or turkey: Mix with chopped avocado, bacon, blue cheese, spinach leaves and tomatoes.
  Corned beef and cabbage: Use the cabbage leaf as the wrap. Spread on some mustard and hot corned beef.
  LBLT: Add strips of bacon and sliced tomatoes, wrap in lettuce and hold the bread!
  Salmon: Mix with chopped avocado, cucumber, sun-dried tomatoes and plain yogurt.
  Tuna salad: Mix tuna with chopped avocado, grated carrots, cucumbers, tomatoes and plain yogurt.

MORE LUNCH IDEAS
- Avocado halves stuffed with egg salad
- Can of sardines

- Ceviche (search 'halibut, snapper and shrimp ceviche' on marksdailyapple.com)
- Homemade soup (search 'perfect primal soup' on marksdaily apple.com)
- Leftover steak topped with salsa and avocado
- Sauerkraut with nitrate-free sausage
- Steamed mussels

- **From the *Primal Blueprint Quick and Easy Meals* Cookbook:** Dill and caper salmon burgers; Orange olive chicken; Pork and shiitake lettuce cups

DINNER – PRIMAL STEAK AND VEGETABLES

At the end of the day, there's no need to make life complicated. A good steak (or chicken, turkey or fish) needs nothing more than salt and pepper and a little bit of time in a frying pan or on a grill to be an amazing meal. Cook some veggies to serve on the side and dinner is served. On the nights when it's convenient to devote more energy and ingredients, take your meat and veggies to a higher level by playing around with new flavours and preparation methods.

1 rib-eye, T-bone or skirt steak, 115–230 g (4–8¼ oz)
½–1 teaspoon salt
½–1 teaspoon black pepper
60 ml (2 fl oz) olive oil
1 onion, sliced
Large handful of mushrooms, sliced
Bunch of kale or spinach, chopped
2 tablespoons of butter
1 tablespoons of chopped parsley

- **Preparation:** Pat dry and season the steak on both sides with salt and pepper. Preheat the oven to 230°C/450°F/Gas 8. Heat a little olive oil in a frying pan over a medium heat and add the onion and mushrooms. Sauté until the veggies have softened and browned slightly, then add the kale or spinach. Sauté until the greens are wilted, then add salt and pepper to taste. Drizzle a little oil into an ovenproof pan (cast iron works great) and then heat the pan on the hob over a high heat for several minutes until it just barely starts to smoke. Drop the steak in the pan and let it sit (don't touch it!) for 3 minutes. Don't be alarmed if there is some smoke. If the steak is stuck to the pan, it's not done browning yet and needs a little more time. If it comes up relatively easily after 3 minutes, flip the steak. Put the pan, with the steak in it, in the oven. Let it bake for several minutes, then check the temperature/texture of the meat to see if it's done (rare 54°C/130°F; medium, 60°C/140°F). Let the meat rest for 5–10 minutes before cutting into it. While the steak is resting, mash the butter with the parsley. Top the steak and vegetables with the herb butter, letting it melt on top, and serve.

- **Primal Steak and Vegetables Variations:**
  Coconut milk: Simmer coconut milk with curry powder until reduced by half, then pour over the meat and vegetables for a quick sauce. Try adding coconut milk to a marinade.
  Cooking technique: Grill the steak and vegetables, or steam the veggies.
  Flavoured Butter: Use your imagination to come up with your favourite flavoured butter. Some flavours to mash into the butter: basil, chives, chilli flakes, cinnamon, blue cheese and bacon bits.

Marinades: Marinate the meat in oil, garlic and fresh herbs before cooking.

Seasonings: Use spices such as cumin, allspice and paprika to flavour the steak.

Vegetables: Rotate different types of vegetables throughout the week (pak choi, broccoli, aubergine, courgette, peppers). Serve mashed or grated steamed cauliflower on the side.

DINNER – PRIMAL MEAT AND VEGETABLE STIR-FRY

340 g (12 oz) of boneless beef sirloin, or chicken, or fish, sliced
*   6 mm (¼ inch) thick*
2 tablespoons (30 ml) sesame oil
4 garlic cloves, chopped
3 tablespoons tamari or low-sodium soy sauce
2 tablespoons olive oil
1 head of broccoli, broken into florets
1 red pepper, thinly sliced
2.5 cm (1 inch) of ginger root, peeled and sliced into thin circles
250 ml (8 fl oz) coconut milk
a few spring onions, chopped

- **Preparation:** Heat sesame oil in a large frying pan or wok on high heat. When the oil is hot, add meat and half of the garlic. Stir-fry one minute for beef and fish, or three minutes for chicken. Add tamari or soy sauce and stir-fry one minute more for beef and fish, or three minutes for chicken. Transfer the meat and its juices to a serving dish and set aside. Heat the olive oil in the same frying pan or wok. Add broccoli, red pepper, ginger and remaining garlic. Stir-fry for 1–2 minutes then add coconut milk. Reduce heat to medium and simmer until broccoli is tender. Combine with meat, and top with spring onions.

MORE DINNER IDEAS

- Breakfast for dinner (omelettes and scrambles)
- Lamb (search 'lamb skewers' on marksdailyapple.com)
- Meatloaf or meatballs with tomato sauce
- Pork chops and sautéed greens
- Pot roast with roasted root vegetables
- Roasted or grilled chicken with salad
- Shrimp sautéed with vegetables (search 'creamy macadamia shrimp' on marksdailyapple.com)

- **From the *Primal Blueprint Quick and Easy Meals* Cookbook:** Chicken and artichokes with garlic sauce; Fish tacos with citrus dressing; Pork fried cauliflower rice

SPRINT WORKOUT SUGGESTIONS

NOVICE SPRINT WORKOUT

Start with five minutes of easy cardiovascular exercise for warmup. Commence the dynamic stretching sequence described on page 171, then 6 x 50-metre warmup 'strides' at moderate effort. Focus on maintaining an efficient stride pattern, with a 20-second rest period between strides. After your warmup, stretching and strides, commence main set:

- 6 x 50-metre sprints of 8–15 seconds each at your estimated 80 per cent of maximum effort. Take a one-minute rest period between sprints, or otherwise rest enough to fully recover and achieve normal respiration before beginning your next sprint.

Use a moving start (jog up to starting line and then begin sprinting) instead of a static start to minimise injury risk. Pay attention to the difference between leg fatigue and pain. If you experience any acute pain or tightness, particularly in your hamstrings, stop the workout immediately with an easy cool down and apply ice to the offended area as soon as possible (15-minute segments of icing, repeated an hour or so apart throughout the day, is the best treatment for an acute strain). After two to three sessions over a few weeks, consider bumping up to the Intermediate Sprint Workouts.

INTERMEDIATE SPRINT WORKOUT 1

Complete warmup, dynamic stretching and 6 x 50-metre warmup strides as described previously, followed by main set:

- 6 x 50-metre sprints of 15 seconds at 90 per cent of maximum effort; 30–60-second recovery interval.

INTERMEDIATE SPRINT WORKOUT 2 – HILL REPEATS

Sprinting uphill is a great workout that stimulates slightly different muscle groups than flat running, and carries far less impact trauma. Complete warmup, dynamic stretching, and 6 x 50-metre warmup strides as described previously, followed by main set:

- 6–8 x hill sprints of 8–30 seconds. Recover by walking or trotting down the hill within 30–60 seconds.

Choose the duration (e.g. 10, 20 or 30 seconds) that you will run on each sprint. You should finish near the same spot at every effort, of course expending more effort to do so on the last one than the first. You don't want to disperse too much energy in early sprints and then fall apart during the later sprints. When you repeat the exact same workout in the future, see if you can finish (on average) further up the hill in 20 seconds (or whatever duration you choose).

INTERMEDIATE SPRINT WORKOUT 3 – ACCELERATIONS

Find a course that you can divide into thirds. You can use a running track and do 150-metre repeats – accelerating every 50 metres (start at turn apex, accelerate at straightaway, accelerate again at middle of straightaway, finish at traditional finish line), or any other venue that you can approximate three equal segments. Complete warmup, dynamic stretching and 6 x 50-metre warmup strides as described previously, followed by main set:

- 6–8 x 30 second sprints with the first third at medium effort, second third seconds at hard effort and final third at full sprint. One-minute recovery between efforts.

ADVANCED SPRINT WORKOUT 1 – SPRINT DRILLS

Complete warmup, dynamic stretching and 6 x 50-metre warmup strides as described previously, followed by main set:

- 2 x 50 metre hopping drill. Drive one knee as high as you can (try to hit your chest), taking off and landing on opposite leg. Then launch off, and land, on the opposite leg, driving other knee high into chest – like an exaggerated skip. It's the same as the warmup exercise on page 172, except you make a maximum effort for height and distance on each takeoff. Fifteen second rest between efforts. One-minute rest before next exercise.
- 2 x 50 metre bounding drill: Take long, exaggerated strides, bounding forwards as far as possible. Unlike hopping, you land on alternating feet as you would running normally. Focus on keeping your balance as you exaggerate a normal running stride. Fifteen-second rest between efforts. One-minute rest before next exercise.
- 2 x 50 metre bunny hops: Take off on both legs and jump up and forwards. Focus on achieving a good balance between height and length. Dip down into squat position for each takeoff, with thighs parallel to the ground. Swing arms skyward to assist takeoff effort and ensure a balanced landing. Thirty-second rest between efforts (you'll need it, trust me!) Two-minute rest before next exercise.
- 4 x 50 metre full speed sprint. One-minute rest between efforts.

HOW TO PLAY

In case it's been a while since you relaxed and frolicked in the outdoors without an agenda, here are some suggested play endeavours:

- **Mini-adventure race:** Bike to the lake, swim to the opposite shore, hike around the perimeter and bike back home. Throw in a skateboard, scooter, or – if you have winter conditions – snowshoes, cross-country skis or ice skates. City-dwellers, try this: hike two blocks to a building with 20-plus storeys and accessible stairs. Climb and descend the staircase, then hike another two blocks to a new building and repeat.

- **Mini-venue sports:** Play a familiar game with altered equipment and/or improvised venues: football on a tennis court or indoor gym; basketball using a hoop or office rubbish bin; volleyball using a badminton net with a beach ball; rounders or tennis in the garden, and so on.

- **Parent competition:** If you're a football mum or rugby dad sitting on the sidelines or driving the team bus, challenge the kids to a friendly competition! Tailor your competitiveness to the size and ability of the kids.

- **Photo scavenger hunt:** A great choice for a landmark birthday party or stag/hen party. A game director is required to administrate – it's just as much fun as playing! Form several teams (the more, the better) of two or three people, each team armed with a digital camera. The game director prepares a list of photographs with corresponding point values based on degree of difficulty. On the photo list, describe points of interest in your area, riddles that reveal a specific location, and outrageous, difficult-to-orchestrate situations (e.g. photo getting shampooed at a hair salon; photo eating watermelon with a stranger on a bus bench; photo astride a Honda Gold Wing motorcycle; photo submerged in a pool holding a bag of crisps – and so on). Distribute the photo lists to the teams at the start/finish venue, start the clock and establish a return time of 2–3 hours. Upon return, each team's photos are evaluated, points tabulated and a winner declared. Print the photos on to a collage for each team for a great souvenir.

- **Speed golf:** Wait until twilight and tee off on an empty course. Carry a junior golf bag with only six clubs. Jog from shot to shot and play quickly, but at the same time making your best effort to score well, including putting. Count one point for each stroke and one point for each minute on the course to produce a total. World record holder Jay Larsen once shot a 71 on a regulation-length (5,500 metres/6,000 yards plus) course in 37 minutes, for a Speed Golf total of 108!

- **Ultimate Frisbee:** My personal favourite! It's suitable for all ages and ability levels to play safely together, with minimal equipment or logistics, in groups of varied numbers. I recommend a minimum of 6 players and a maximum of 16. Depending on the size of the group, a field of 45–90 metres (50–100 yards) length and 32–45 metres (35–50 yards) width is ideal. The game (the proper term is simply 'Ultimate', since Frisbee is a brand name) is somewhat like football with a flying disc. Teams try to score a goal by covering the length of the field passing the disc and crossing the end line. You can pass forwards or backwards, but cannot run once the disc is caught. If a team drops the disc, the other team takes possession immediately on the spot of the drop and tries to pass to open players and get across the opposite end line – it's non-stop action! Players should match up with opponents appropriately by size and ability, covering players from the opposing team in man-to-man defence style. However, no physical contact with an opponent is allowed except incidental contact going for the disc. These rules allow for the full and safe inclusion of a diverse group.

INDEX